The Hen Harrier's Year

THE HEN HARRIER'S YEAR

Ian Carter and Dan Powell

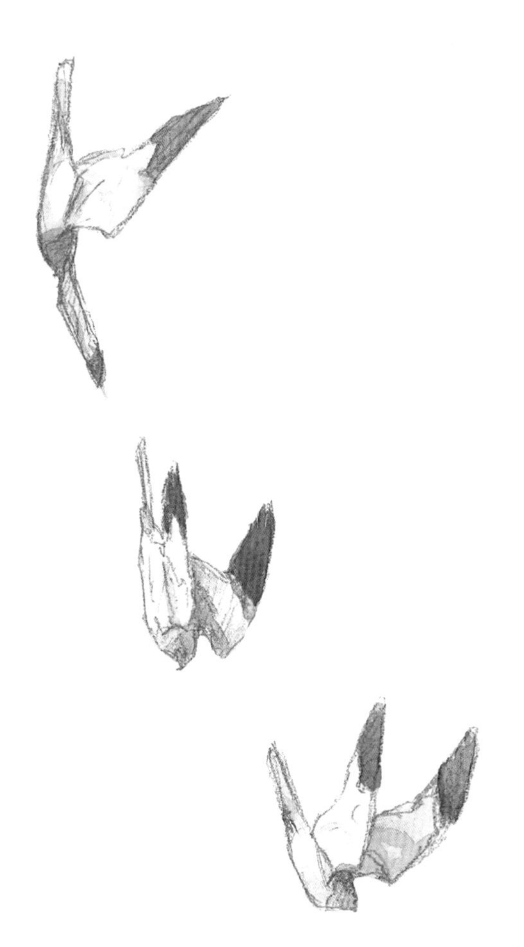

PELAGIC PUBLISHING

Published by Pelagic Publishing
20–22 Wenlock Road
London N1 7GU, UK

www.pelagicpublishing.com

The Hen Harrier's Year

ISBN 978-1-78427-385-9 *Paperback*
ISBN 978-1-78427-386-6 *ePub*
ISBN 978-1-78427-384-2 *PDF*

https://doi.org/10.53061/OPFD3750

A CIP record for this book is available from the British Library

Cover: A male takes over guard duty as the female returns to the nest

To the watchers on the hill

CONTENTS

THE SHADOW CHANGED
SHAPE SHAPES
AS IT PASSED
DIFFERENT
21/5/19 AM.
SHADOWS & LIGHT
AT TIMES IT WAS EASIER TO FOLLOW THE H. BIRD'S SHADOW
HUNT
THE

FOREWORD

Most birdwatchers in Britain have their first encounter with a Hen Harrier at one of the bird's lowland winter roosts. In the early 1980s, Hen Harrier was a regular highlight of my midwinter visits to the Lincolnshire coast during weekends and school holidays, ample reward for an hour's bike ride on a bleak December or January day. A male in adult plumage is one of our most eye-catching birds, thanks to a combination of its striking plumage and elegant, effortless flight. Coastal areas are occupied only in the coldest months and the birds disappear as late winter gives way to early spring, heading back to the remote upland areas where they breed.

In recent years the Hen Harrier has become the de facto flagship species for the birding community in its stance against raptor persecution. That has brought our focus on the Hen Harrier firmly to those upland breeding grounds, vast areas of heather moorland which it shares with other raptors and a handful of other species, including the Red Grouse. The grouse-shooting industry, and specifically the pressure for gamekeepers to generate a large surplus of birds for the bewildering practice of driven shooting, has led to conflict – harriers will take grouse chicks and can have an impact on grouse numbers. Persecution is illegal, but policing is easier said than done. It is not just Hen Harriers that are affected, but they have suffered more than most, with the result that the species came close to extinction in England, while numbers in Scotland and Wales are in serious decline. The extraordinary skydancing display of the males above their breeding territory in spring could become something that future generations of birdwatchers in this country can only read about.

This book tells the story of the Hen Harrier throughout the year, tracking the behaviour and ecology of this stunning bird from its winter roosts, through the breeding season and back again to winter. Further chapters explore its world status, history in Britain and of course the harrier/grouse conflict already mentioned. The two authors are well equipped for the job. Ian Carter has a lifelong interest in and professional experience of Britain's raptors; while he is best known for his work on the Red Kite, he has a deep knowledge of the Hen Harrier too. Dan Powell's illustrations have brought so many books and articles to life, and his ability to capture both movement and plumage is showcased throughout. These two have drawn on their own knowledge and experience, but also the research of others, and they have made good use of the papers and reports published in *British Birds*, the journal through which I came to know them both, and which has charted the recent developments in the raptor persecution story through its news pages.

This is a particularly timely book, as we survey the Hen Harrier's current predicament in Britain. The bird needs our help – and I am certain that this book will make a significant contribution to a wider understanding of the issues involved.

Roger Riddington (former Editor of British Birds*), Shetland*

ACKNOWLEDGEMENTS

It was a privilege to work with colleagues involved with Natural England's Hen Harrier Recovery Project over many years, in particular Andy Brown, Allan Drewitt, Phil Grice, Matt Heydon, Adrian Jowitt, Stephen Murphy, Richard Saunders and Nigel Shelton. Stephen Murphy took me to my first Hen Harrier nest in the Bowland Fells many years ago and that experience, together with further field visits (and his boundless enthusiasm) encouraged my interest in the bird and, ultimately, led to this book. Simon Lee at Natural England provided useful information about the proposed lowland reintroduction.

Guy Shorrock supplied details of incidents of illegal persecution on the RSPB's reserve at Geltsdale and elsewhere. He and his colleagues (and ex-colleagues) in the RSPB's Investigations team, including Bob Elliot, Duncan McNiven, James Leonard and Mark Thomas, have worked tirelessly, in the most difficult of circumstances, to highlight the routine illegal killing that sadly still takes place on our grouse moors.

I'm grateful to Neil Calbrade at the British Trust for Ornithology for permission to reproduce the distribution maps for Britain & Ireland. The late Roger Clarke was a pioneer in identifying prey from the remains of raptor pellets and became a world authority on the subject. I had useful discussions with him about raptor diet and communal winter roost sites over the years through a shared interest in the subject. While the focus was on our joint work on the Red Kite, it was always apparent that harriers were his favourite birds and they were rarely far from the conversation.

It may seem odd to thank two people I've never met, but I feel I have, in some small way, come to know them both through writing this book. Eddie Balfour worked for RSPB on Orkney and was involved in pioneering studies of the Hen Harrier in its stronghold on the islands for over 40 years, until his untimely death in 1974. He described the widespread polygynous breeding system that is prevalent in Orkney and established a baseline of information that others built on, both in Orkney and on mainland Britain as the bird recolonised. Donald Watson was another pioneer, studying harriers at countless nest sites and communal winter roosts close to his home in Dumfries & Galloway. His book on the Hen Harrier, published in 1977, combines inspirational writing and his evocative artwork, and is regarded as a classic of its kind. My old, battered copy is as well thumbed as any book I own.

The good folk at Pelagic Publishing were as helpful as ever in guiding this book to completion. David Hawkins and Gillian Bourn provided invaluable editorial expertise. Nigel Massen oversaw the project with patience and helpful guidance from beginning to end, and Rhiannon Robins has worked hard to try to get it into the hands of as many readers as possible. BBR Design have used their typographic skills and design mastery to build the book you see before you.

Finally, my wife Hazel endured long trips to harrier roosts on Dartmoor in some of the worst conditions that midwinter has to offer – and even caught sight of a harrier or two.

Ian Carter, Mine House, Galloway

A huge thank you to Steve Downing, RSPB and Stephen Murphy, Natural England for their immediate enthusiasm and encouragement when first I explained to them about the project. It made a huge difference having their unconditional trust and confidence in what we were trying to achieve with this book. Additionally, I must thank them for allowing me to bend their ears on the phone on all things harrier – this also goes to John Miles and Colin Shawyer. Colin provided me with references for the tree nesting Hen Harrier and Merlin drawings.

Cheers to the 'folk hidden in the heather' who watch over the moors – their dedication goes above and beyond. We met and chatted with a couple of them: Robert Matthews and Sel. Another was tucked away among the rocks – on reflection, we really must be more selective as to where we nip off to when nature calls. Thanks to Mike Price for sharing his knowledge, giving up his time to lead us to a safe spot to watch from and giving my heart a good work out on the way there. Pat Martin, another champion for the harriers – great company, unselfish sharer of his experiences, photo references and 'Rosie' namer.

The late David Cobham, whom I got to know a little while working with him on *Bowland Beth* (see *Conflict on the grouse moors*). Thank you for planting the seed that grew into a deep emotional connection with these critters. I miss our long conversations, especially the sweary ones about drawing foxes. The Hark to Bounty Inn, where we had dinner and raised a glass in honour of David and Beth. The friendly hospitality at The Shoulder of Mutton. Dobry Domuschiev and family, great friends and for sharing their Bulgaria with us. Pavel Simeonov maker of eye-melting brandy and kind host on the Black Sea Coast. Nigel Massen for letting us loose on the sister book to *The Red Kite's Year*. Last, but definitely not least, my wife Rosie. Thank you for giving your permission to include your lovely sketches made during our trips up north (initialled RP). They add tremendously to the story of the moorlands. Where to next?

Dan Powell, Hillhead, Hampshire

INTRODUCTION

The charismatic Hen Harrier can brighten the mood of any birdwatcher. It has that special blend of qualities which makes it stand out from the crowd. It is scarce in Britain but just common enough to make seeing one a realistic possibility if you spend time in the right places. The pale grey males have an ethereal beauty and (unusually for birds of prey) look very different from the females and young. For a long time, early ornithologists considered male and female/juvenile Hen Harriers to be two separate species. Perhaps best of all, unless you are visiting an established breeding area or a winter roost site, the bird's appearances are unpredictable and often take you by surprise. When we lived in a farmhouse in the arable prairies of Cambridgeshire it was a species I kept a special eye out for in winter. Weeks would pass without a sighting and then, when I'd all but given up hope, a stunning grey male would suddenly materialise, Skylarks spilling away to the side and out of harm's way.

The Hen Harrier can be rather understated in its behaviour: long glides a few metres above the ground on stiff, slightly raised wings are interspersed with a few short flaps, as it relentlessly scans the vegetation below for food. But it is capable of the spectacular as well. In the breeding season, the males indulge in the most remarkable 'skydancing' display flights to warn away rival males and impress the nearby females, diving bullet-like towards the ground before pulling up dramatically, and then repeating the performance, sometimes dozens of times in quick succession. In winter, too, it provides a memorable spectacle at traditional sites where birds gather in numbers in the late afternoon gloom to roost together on the ground.

Its preference for big landscapes and wide, open spaces means you often have to settle for a distant view. But occasionally, usually when it is furthest from your mind, a Hen Harrier will come gliding gracefully by within a few metres, drifting on towards the horizon as you try to catch your breath. The fact that this species is under great pressure in Britain adds a touch of melancholy to each sighting:

you are glad your bird has managed to survive thus far to hunt the local fields but, at the same time, you are worried for its future.

Of the three species of harrier that regularly breed in Britain, the Hen Harrier is the most common and widespread, although the larger and more robust Marsh Harrier is catching up rapidly. The elegant Montagu's Harrier is the smallest and least common, with just a handful of breeding birds in most years. It is similar enough to the Hen Harrier to present an identification challenge. Adult males of both species are predominantly pale grey and the females share a very similar, intricately patterned brown plumage. The Montagu's Harrier is a more delicate bird, with proportionately longer, thinner and more pointed wings, and it is only with us in the summer, before it heads back south to wintering grounds in Africa. If you have spent time watching these birds, the differences between them are clear enough, but an isolated individual, especially a female or young bird, can prove tricky if you are out of practice.

There is a third harrier in which the male is predominantly pale grey: the ghostly and highly sought-after Pallid Harrier. This is an Asian and eastern European species but it has been spreading gradually westwards and is becoming more frequent in Britain. It breeds regularly in Finland and there have been recent breeding attempts in arable farmland in Spain and the Netherlands. In Britain, males have occasionally paired up with female Hen Harriers and attempted to breed, though, as yet, not successfully – hopefully it is only a matter of time. These two birds have a close affinity and hybrids are occasionally seen in northern and eastern Europe, causing identification pitfalls for the unwary.

The only other harrier to have occurred in Britain is the Northern Harrier of North America, sometimes referred to as the Marsh Hawk. Individuals have strayed across the Atlantic on at least nine occasions and have no doubt been overlooked at other times due to their similarity to the Hen Harrier. Indeed, these two birds were once considered to be a single species. Most authorities now consider them to be separate and there are clear, if rather subtle, differences in plumage between the two birds. The alternative name Marsh Hawk is an appropriate one for the American bird, alluding to its habitat preferences. In contrast to the Hen Harrier, it favours wetlands in the breeding season, making full use of this habitat in the absence of a North American equivalent of the Marsh Harrier. In addition to the five species already mentioned, a further 8–11 species (depending on your choice of taxonomist) complete the harrier clan, occupying various parts of South America, Africa and Australasia. All are broadly similar in size and share the typical harrier characteristics.

The Hen Harrier's scientific name is Circus cyaneus, *the Latin* 'cyaneus' *meaning blue-grey or dark blue – a clear reference to the plumage of the male.*

What is it that sets harriers apart from other birds of prey? Perhaps it is their mode of hunting that most clearly defines them. Harriers are lightweight, buoyant birds that travel low to the ground with their wings in a distinctive, stiffly held V-shape when gliding, as they scan the vegetation below for prey. They are adept at sustaining flight for long periods and so are able to cover huge distances, allowing them to exploit tracts of open country with few perching opportunities.

The generic name *Circus* is derived from an ancient Roman word for an oval arena or race-track, and means 'the hawk that circles'. It may allude to the wide, roaming foraging flights or, perhaps more likely, refers to the effortless circling high above the breeding grounds in spring. As expected for a bird of open country, harriers typically build their nests on the ground, despite the threat to eggs and chicks from roving mammalian predators. There is one species, the Spotted Harrier from Australia, that nests in trees, and in Northern Ireland the Hen Harrier itself has occasionally adopted this tactic in forests planted within its moorland breeding areas.

Harriers have a fringe of stiff feathers that forms a disc around the edge of the face. This is more typically seen in the owls, and helps to focus sounds so that prey can be pinpointed more accurately. It allows nocturnal owls to drop onto

The distinctive harrier facial disc is evident when a perched bird is seen close-up, and may be noticeable in flight, particularly if a bird is heading directly towards the viewer. The ruff of feathers can be raised to expand the width of the facial disc which helps improve the way that sound is focused into the ears – in humans the cupping of hands behind the ears has a similar effect. As with owls, harriers often keep their heads pointing vertically downwards towards the ground to maximise their ability to hear, as well as to see, their prey.

prey that has been located purely by the sound it makes, and it is no doubt useful for harriers when hunting small mammals concealed by vegetation.

The other trait that sets harriers apart and which is much admired by all who watch them at their breeding sites is the food pass. In birds of prey it is often the male that shoulders the greatest burden in finding food during the breeding season, leaving the larger female to incubate eggs or guard chicks in the nest. When food is brought in it must be transferred to the female so that she can feed herself, or take it back to the nest to feed the young. In many species this is achieved using a handy branch or other perch close to the nest. But in the open country used by harriers that is not an option. Instead, food is transferred in mid-air, a unique behaviour that is explored more fully in the chapter for *June*.

Not all harriers are polygynous, but this mating system is seen regularly in the Hen Harrier. Males mate with up to seven different females though such a high number is exceptional. In contrast, most raptors are generally monogamous, excepting a few mischievous extra-pair copulations. Polygyny appears to be another harrier trait associated with the open country in which it breeds. Only in an open landscape with good sightlines can a male hope to keep tabs on several breeding females to try to make sure that any young produced are his rather than those of a rival male.

The Hen Harrier occupies a vast range across Europe and Asia, extending from the Atlantic coasts of Western Europe, across to the Pacific coastline of eastern Russia. In most of its range this species is a true migrant, moving well to the south in order to escape harsh winter conditions and snow cover in its northern breeding areas. There are only a few areas, including Britain and Ireland, where the relatively benign maritime climate allows it to remain year-round, but even here, there are dispersive movements. The use of high-tech radio and satellite tags has considerably improved our understanding of the complex movements that are made by our 'resident' birds.

As with many other raptors, the Hen Harrier has a chequered history in Britain. The first written record of the bird, by William Turner in 1544, highlighted the problem: 'it gets [its] name among our country-men from butchering their fowls'. Concerns for domestic hens and gamebirds meant that it was pushed to the edge of extinction at the hands of humans towards the end of the nineteenth century, before staging something of a comeback. Many raptors have recovered in recent decades but the Hen Harrier remains under extreme pressure from humans today. In fact, it is centre stage in one of the most difficult and intractable conflicts between elements of the shooting community and wildlife. It is viewed as a threat to the highly esteemed Red Grouse, valued for the lucrative shooting it provides across large areas of our uplands. Persecution by humans and the so far unsuccessful attempts to stop it are, sadly, a recurring theme throughout this book. Whereas the subject of the previous book in this series, the Red Kite, has a bright and secure future ahead of it, the same is far from certain for the beleaguered Hen Harrier. Its fate hangs in the balance, dependent on intensive conservation efforts and, most importantly, our attitudes towards it.

The harriers are mid-sized raptors, larger than most hawks and falcons, but smaller than the eagles and vultures. The Hen Harrier (above: adult male top left, sub-adult male centre, albino male bottom left, females/'ringtails' to the right) is slightly larger and more robust than its close relatives the Montagu's Harrier (right-hand page, upper two birds) and the beautiful Pallid Harrier, the palest of the three (right-hand page, middle). The structure of the wings is subtly different; the Hen Harrier has the broadest wing-tips and the Pallid Harrier the most pointed, with just four visible 'fingers'. The male Hen Harrier lacks the Montagu's Harrier's black lines across the wings, and yet it was not until the early nineteenth century that they were properly established as two separate species – by George Montagu who gave the smaller bird its name. Finally, the distinctive Marsh Harrier (male, bottom right; female, bottom left) is the largest of the four species.

The term 'ringtail', applied to Hen, Montagu's and Pallid Harriers sometimes causes confusion. It refers to the predominantly brown-plumaged adult females and young birds of either sex. The name comes from the alternating bands of dark and pale along the tail, and the narrow but obvious band of white across the rump, at the base of the tail. This last feature is often visible at surprisingly long range and sometimes, especially on a dull day, it almost glows out from the gloom, providing a point of brightness on a brown bird that otherwise tends to blend into the background.

Another issue of *British Birds* magazine tells of the disappearance of two more Hen Harriers from the moors. It's shocking but it's not news: virtually every issue that drops through our door carries similar reports of raptor persecution. The regularity of the reporting almost leaves me numbed into accepting this as the norm. What can I do about it? When the opportunity arose to work again with Ian, this time on Hen Harriers, it would have been easy to slip into rant mode throughout. However, the world is angry at the moment, there are plenty of rants over raptor persecution and yet the organised slaughter continues without reprisals. Ranting isn't working.

There are other ways to fight their corner. Mine is to celebrate the lives of these wonderful creatures, maybe pricking at a few closed minds along the way ... oops, tiny rant.

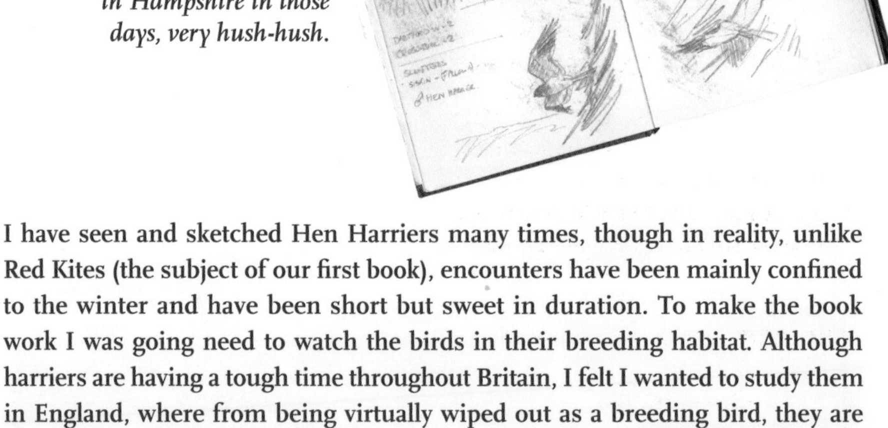

Early sketches of a New Forest bird. Nice to see a Goshawk too – a rare bird in Hampshire in those days, very hush-hush.

I have seen and sketched Hen Harriers many times, though in reality, unlike Red Kites (the subject of our first book), encounters have been mainly confined to the winter and have been short but sweet in duration. To make the book work I was going need to watch the birds in their breeding habitat. Although harriers are having a tough time throughout Britain, I felt I wanted to study them in England, where from being virtually wiped out as a breeding bird, they are showing tentative signs of recovery. Out of respect for the trust given to me by those who supplied the locations of nesting birds, I have not named any of the areas we visited during the summer of 2019.

Hen Harriers are another artist's bird. When not vanishing into the distance, they may bide a while, allowing time to observe the beauty of the shapes they form while hunting: tails twisting-twitching, wings flicking, the head held steady, eyes alert for food, turning and plunging into the heather. The male in particular, with his striking grey, white and black plumage, is irresistible to the

Watching and waiting at a traditional winter roost site can afford an opportunity to observe Hen Harriers at length – it doesn't always work out that way. New Forest, 1990s. A herd of Fallow Deer grazing on the plain provide further interest to the scene – the idea on the right and below later developed into the studio watercolour above.

artist's brush. Almost all the images in this book have evolved out of sketches and paintings made in the field. Some were completed on site, others finished back in the studio and a few reworked after being retrieved from the bin. Although one or two were produced especially to go with Ian's text, the majority were already completed and tied in nicely with his words.

The 1970s and 1980s were my apprentice art and birding years. Most time was invested at Titchfield Haven National Nature Reserve, a fabulous site for getting to grips with wildlife and nature conservation management. Every now and then a Hen Harrier would pass through or stay a while. My memories of these moments were of watching in awe as a ringtail landed in front of the hide, staring at me (maybe working out if I was edible), then flying from the post, gliding low over the meadow, using the cattle as cover to flush out some lunch. There was another magical moment during the cold snap of January 1987, when a dusk watch was rewarded with a male and ringtail Hen Harrier coming to roost in the reedbeds, joined by a Barn Owl and Bittern. I also remember the rather dodgy ride home on my old Honda trials motorbike along dark, snow-covered lanes.

During our time with the nesting birds we resolved to take a neutral, hard-nosed approach to our observations. Like that was going to work! From the first encounter with a male, we became fully invested devotees of Hen Harriers and their world. These encounters can evoke deep emotional connections in people; however, nature is completely oblivious to this fact. If harriers feel any emotion, it would certainly not be directed towards us. Maybe watching these birds with this in mind, as they instinctively try to get on with their lives while 'we' are doing our best to bugger-up everything for them, made our bond with them even stronger.

Throughout our time with the harriers we wanted to try to build an overall picture of the other species that make up the biodiversity of the upland moors. Rosie noted: While the birds were out of view, off doing what they do, I took inspiration from the flowers and insects all around us. Top left – Thyme-leaved Speedwell, Lesser Spearwort, Climbing Corydalis and Heath Bedstraw (RP). Centre – Ring Ouzel. Top right – Dipper. Bottom right – Mountain Pansy, Cranberry, Bog Pimpernel, Globeflower and Bilberry (RP). Above – the weather was better on some days than others, but the birds were still active no matter what was thrown at them – they had to be.

ROUZEL 21/5/14
DIPPER - BUSY WORKING THE RIVER DARTMEET
1230 29/1/60

- SILVER WINGS ACROSS

JANUARY

THE SHORTEST DAY HAS PASSED but daylight remains a scarce resource during the first month of the year. Time available for hunting is limited. January is often our coldest month and in cold weather more food than usual is required to maintain condition and prevent body temperatures from dropping too low. If there is heavy snowfall, food becomes harder to find as small mammals hide beneath the sheltering blanket that now covers the ground. To add to the challenges, the amount of food available is on a downward trend, as the cruel realities of winter begin to bite. Small mammal and bird populations are in decline as the weakest succumb to the cold, and predators take their toll. The replenishing breeding period is still many months away. If there's a silver lining for the Hen Harrier it's that these challenging conditions help to make avian prey a little easier to catch. In tough times, prey species must sacrifice some of their usual caution and vigilance when trying to keep themselves well fed. If finding enough food to stay alive is their primary concern, they may venture further from cover, with a less watchful eye than usual on the skies above.

Communal roosting

In the depths of winter, the lives of most Hen Harriers are centred on one or more communal roost sites where they spend each night with others of their own kind. In Britain, roosts tend to be small, sometimes just a handful of birds, particularly in parts of the country where harriers are scarce. There are simply not enough birds within reasonable commuting distance to support larger gatherings. But in favoured areas, roosts on mainland Britain can, exceptionally, support as many as 30 birds, and on the Continent roosts of 50–60 birds have been reported. The largest roost gatherings recorded in Britain and Ireland, and probably the whole of Europe, are on the Isle of Man. Here, the most important roost once supported no fewer than 160 birds, a substantial proportion of the island's Hen Harrier population. Sadly, numbers have fallen drastically in recent years, in part due to disturbance by introduced Red-necked Wallabies.

As befits a bird that evolved in open, treeless country, communal roosting takes place within rank vegetation on the ground. A wide variety of different habitats are used, but they all share a number of key features. The ground is usually flat or gently sloping and the site tends to be free from regular human disturbance. Dense, reasonably tall vegetation in a more or less sheltered location is required, with small open patches where the birds can rest, concealed and protected from the elements by the vegetation around them. Bracken, reeds, sedges, grasses, rushes and heather are all used if these basic requirements are met.

Waterlogged sites are often exploited, and probably help to reduce the risk from ground predators, particularly Foxes, as well as discouraging human visitors. Within such sites the birds rest on trampled, flattened patches of vegetation that provide a dry spot, allowing them to pass the night without immersing

New Forest roosts. A different colour palette for each evening.

their feet in water. The renowned Hen Harrier artist Donald Watson referred to these patches as 'beds' or 'forms', pinching the latter word from the resting places of the Brown Hare, another animal of open country. He was able to find them relatively easily when visiting roosts in daylight from the appearance of the trampled vegetation and the splashes of white droppings.

Trees are present at some sites, although they do not appear to be used as perches. Patches of heather or grass along wide rides within forests may be used,

Stunning close-up views in a spectacular setting as the light fades into evening. Some roosts make all those invested hours and numb extremities worthwhile.

and the once huge roost on the Isle of Man included scrubby willow and birch trees spread over an extensive area. The birds would no doubt be better protected from ground predators if they took advantage of trees where they are available, but evolution proceeds slowly and this bird has long been associated with open country. The likely benefits for harriers of communal roosting are considered in the *October* chapter; the pleasure to be had by humans when visiting a roost to watch them is described in *December*.

Distribution in winter

In northern and eastern Europe, and in Asia, the Hen Harrier is a genuine migrant. It heads south in the autumn, vacating the breeding areas and moving to separate wintering areas in southern Europe, north Africa, India and southern Asia. In Britain, the story is more complicated. The upland breeding grounds are not entirely vacated and harriers may be seen there throughout the winter, though they tend to move away from the highest and least hospitable sites. The onset of severe weather can result in more birds dropping down onto the lower ground, or moving away to nearby coastal sites. South-west Scotland, the Hebrides and Orkney in the far north usually escape the worst of the winter weather and harriers remain here throughout the year. Orkney is the most northerly wintering site for the Hen Harrier anywhere in the world; indeed, no harrier of any species is found further north in winter.

Large parts of southern Britain, well away from the breeding grounds, support Hen Harriers in winter, especially open tracts of land with limited human disturbance. This is the time of year when most birdwatchers have their best chance of catching up with them locally. Coastal and low-lying sites are favoured, with concentrations around estuaries and coastal plains in south-east England and along parts of the south coast. The flatlands of the fens in eastern England support small numbers as does the wide expanse of Salisbury Plain in Wiltshire and other open, downland sites in southern England.

Well inland, across large parts of central England, wintering harriers are few and far between, but even here, suitable expanses of open country such as old airfields or wetland nature reserves may occasionally attract a bird or two. In Ireland, some individuals remain in the breeding areas throughout the year, but the bird becomes far more widespread in winter and may be found widely in lowland and coastal sites where it is absent in the summer.

Because the smaller males are more likely to move away from their upland breeding areas to avoid harsh conditions and track their favoured prey of small songbirds, they occur disproportionately in western and southern wintering areas. Up to 30–45 per cent of birds seen at communal roosts in the south and south-west of England are grey-plumaged adult males. In eastern England the proportion tends to be lower, with roosts more likely to be dominated by ringtails. This may also be influenced by the presence of continental birds, dominated by brown-plumaged juveniles, in eastern areas, though it is thought that the overwhelming majority of birds wintering in England originate from breeding areas within Britain.

The Hen Harrier is widely, if thinly, distributed in Britain and Ireland in winter, although it is scarce across many inland counties, with sightings unpredictable away from known roost sites. (Map from Bird Atlas 2007–11, *a joint project between BTO, BirdWatch Ireland and the Scottish Ornithologists' Club – reproduced with permission from the British Trust for Ornithology.)*

USING THE HEDGE AS COVER

FEBRUARY

THE DAYS CONTINUE TO LENGTHEN in the second month of the year and, on a bright day, with the sun now higher in the sky and the earliest songbirds in full voice, thoughts turn easily towards the coming spring. Yet it can also be the most gruelling of months, with bitter cold and snow, made all the more difficult for wildlife because it comes at the end of winter, when resources are at their lowest ebb. February can be the final nail in the coffin for a Hen Harrier that has, so far, managed to scrape a meagre existence through the coldest months.

Hunting behaviour

The Hen Harrier is an opportunist, ambush predator of small birds and mammals. Its modus operandi, as with other harriers, is to fly along relatively slowly, a few metres above the ground, perpetually scanning the vegetation below. A breeze makes flight more efficient, reducing the amount of energetic flapping required and increasing hunting success. And by flying into the breeze, a harrier can maintain its height while flying at a slower speed, giving it more time to examine the ground below in detail.

Watching a hunting bird for any length of time shows how it varies its speed to increase the chances of a kill. Slow flight, with long glides, is useful to check carefully for prey concealed in vegetation, and bouts of hovering allow it to hold station, briefly, so that potential prey can be scrutinised more closely. But a harrier will gain speed rapidly if it senses an opportunity to ambush birds, having masked its approach behind cover or a fold in the landscape. Low posts or mounds are used by resting birds and no doubt the opportunity is taken to watch for potential prey even when not on the wing. There are also occasional records of birds landing on the ground and waiting for prey to emerge from the surrounding vegetation, before pouncing.

Despite the languid appearance of its low foraging flights, it is easy to imagine that a hungry Hen Harrier is on perpetual high alert as it patrols the winter landscape. An hour might slip by with no signs of prey and then, suddenly, a small bird sees the approach too late and falls within striking range. Many attacks are unsuccessful, so persistence is required to secure a meal.

The large, forward-facing eyes are adept at spotting movement and provide binocular vision so that distances can be judged accurately. Harriers also possess a keen sense of hearing. The distinctive, owl-like facial disc has a stiff fringe of short feathers which helps to focus sound, allowing it to home in on tell-tale rustling noises that indicate potential prey, even when it is concealed by vegetation.

The Hen Harrier is, then, adept at covering the ground and getting itself within striking range. The final piece in the jigsaw is the catching and killing. For this, the bird's long legs are deployed as it strikes down into the vegetation or attempts to grab a bird as it flies up. The sharp talons open and, if the attack is successful, close again over the prey, securing it firmly. Speed and manoeuvrability are crucial to increase the likelihood of success, and the long tail (which can be fanned out) helps to slam the brakes on when required, as well as facilitating rapid changes of direction. More familiar birds, such as the Pied Wagtail, use their tails in much the same way when launching themselves after flies. The Hen Harrier is rarely able to hunt successfully following a prolonged pursuit. To make a kill, everything depends on the initial element of surprise and lightning speed over a short distance. If the prey escapes the initial pounce and perhaps a short chase, it is likely that it will live for another day.

A hunting Hen Harrier will sometimes fly over a likely patch of cover before turning back to investigate more closely. The way it methodically scrutinises the landscape led one French observer to describe a hunting harrier behaving 'as if it were searching for a lost object', and birds do give the impression of working the ground diligently as they check for potential prey.

A harrier flying into the breeze can turn and drop down at speed once it spots something, stalling by fanning out its tail and then using the wind, now in its favour, to assist the dive. Hen Harriers will deliberately fly over flocks of

Danube Delta. Two ringtails had been quartering an area of rough ground by the side of the road with low, focused sweeps. A Meadow Pipit breaks cover and with a powerful turn of speed and agility is dispatched with the flick of a talon.

Meadow Pipits in a spot of bother.
Ashley Hole, New Forest.

birds, such as waterfowl, that are larger than their usual prey. They are probably checking out the flock, just in case there is a sick or injured bird present. If all fly up strongly then the hunter will quickly move on without wasting time attempting to make a kill.

Sightings of Hen Harriers at night, caught in torchlight or by car headlights, offer the intriguing possibility that they may hunt in darkness if food has been hard to come by in the day. They are certainly seen hunting in poor light around dusk and some birds leave their winter roost early in the morning when light levels are still very low. Hunting in darkness is harder to prove, but birds are occasionally seen arriving at roosts well after dark, something that would usually be missed and so may happen more often than suspected. The suggestion of night hunting is perhaps not as far-fetched as it might seem given that harriers have excellent hearing. Indeed, ingenious experiments involving the American Northern Harrier, in the lab as well as in the field, have shown that hidden mammals can be located through sound alone. On occasion, the tiny concealed microphones playing vole sounds were struck and carried away by the hunting harrier.

Foraging range from roost sites

It is no surprise that Hen Harriers range considerable distances from their overnight roost sites when questing for food. This is, after all, a bird of wide, open expanses of countryside, lightweight for its size, and so able to cover large distances with a minimum of effort. In one sense, travelling and hunting are the same thing. As a new day begins and a bird flies out from its roost, skimming low over the vegetation, it has already started its search for sustenance. It may have its mind on a good hunting area some distance away, but will not turn down the chance of a meal should one present itself en route. Much like a migrating Swallow, it can hunt while in transit.

The Hen Harrier has long wings and a lightweight body – a light 'wing-loading' to use the jargon. In favourable conditions this allows a hunting bird to exploit the breeze, gliding over the landscape with just a few wing-flaps required to maintain momentum. This is a highly efficient form of flight, enabling it to cover huge distances in its relentless search for food. In contrast to most raptors, a harrier can spend many hours in the air each day. Individuals have been seen

Bishops Dyke, New Forest, 2013. Sometimes the Forest delivers in bucket loads, other times it doesn't. This was one of those days. A bright day's wander around Bishop's Dyke should have been full of 'Attenborough' moments, except that it wasn't. In truth it turned out to be a sodden-footed traipse. Apart, that is, from five minutes of bliss when a handsome male Hen Harrier graced the heathland – twisting and turning on a sixpence, handling sudden gusts of wind with consummate assurance, in search of prey. Just time for a couple of sketches and all too soon it ended. With a flick of his black wing-tips he slipped away over a heather ridge and was gone. A splendid cameo that made the return home via Ikea almost bearable – actually it was unbearable.

hunting 15–20 kilometres away from their roost, and estimates suggest they may fly a total distance of up to 160 kilometres (the equivalent of about four marathons) in an average day. Harriers are flying-machines, hunting with persistence, patience and no shortage of stamina.

The Hen Harrier is very different from highly territorial birds of prey such as the Common Buzzard or the Kestrel. These species actively defend a small but diverse territory that may be around 1–2 square kilometres, perhaps less. They must find all their food throughout the year within this small area and, unsurprisingly, they will not welcome competition from others of their own kind, and will drive away rival birds. Other birds of prey are not territorial but forage over wide areas as a group, looking for food resources that are shareable. The vultures are the most obvious example. If a large carcass is spotted by one bird, others will quickly move in to take advantage and dozens of individuals may end up with a good meal. Red Kites forage in a similar way, keeping an eye on each other and quickly homing in on a site when food, such as a carcass, is located.

The Hen Harrier's strategy is different again. It is a lone hunter, like the Kestrel and Buzzard, but it roams over great distances in search of prey, like the vultures, sometimes hunting over tens of square kilometres. It may occasionally encounter another Hen Harrier on its travels, but it would not be practical to try to defend such a large area against others of its own kind. In places where there is a good food supply, several harriers may congregate and here territorial behaviour can be seen. Dominant birds, generally the larger females, will drive away other harriers in order to keep the best hunting for themselves. It has even been suggested that the brown plumage of immature birds of both sexes may be a form of mimicry. If youngsters look like the larger adult females, other birds in the area may think twice before challenging them.

Away from food-rich sites and winter roosts, Hen Harrier sightings tend to be unpredictable in winter. If you see a Buzzard or a Kestrel in your local countryside, you'll have a good chance of finding it again in the same area on subsequent days. But if a Hen Harrier cruises by, it might be many kilometres away from its roost site and it may not visit the same field again for several weeks. It is worth making the most of each and every sighting.

The overall distances that Hen Harriers cover when away from their roost sites will vary depending on the food supply in the local countryside. Clearly, it makes little sense to travel vast distances from the roost if high-quality habitat, rich in prey, is to be found within a few kilometres. But in less productive landscapes, or if conditions change and prey dries up, birds will need to explore farther afield.

A Hen Harrier's lifestyle in winter might be described as semi-nomadic. The bird is constrained, to some extent, by the location of communal roost sites, but it is always on the lookout for new hunting opportunities and willing to adjust its routine accordingly. Away from roost sites, you may be lucky and see a bird in the same place several days running if the hunting is good. Or weeks may go by without a sighting.

Winter habitats

Although Hen Harriers are capable of covering huge distances, they don't simply roam around at random. Large areas within their foraging range provide few opportunities for successful hunting. Bare agricultural land such as ploughed fields or closely cropped grasslands lack the vegetation required to allow prey to be ambushed – a bird that can see a harrier's approach will easily be able to stay out of harm's way. Equally, vegetation that is too tall or dense is unsuitable, as a hunting harrier needs to be able to make a clean strike at its prey. Dense, uniform scrub, lacking any gaps, and well-established woodland are generally avoided.

What a harrier needs is something in between the extremes of bare ground and dense, impenetrable cover. It wants sufficient vegetation, or landscape features, to allow it to approach unseen, but not so much as to prevent a strike. These requirements can be met by a variety of different vegetation types and this explains why harriers are able to exploit a wide range of habitats. Farmland, heathland, moorland, rough grassland, saltmarsh and young forestry plantations all provide useful foraging habitats. There are some habitats where the majority of the land is unsuitable but where linear features such as ditches, streams, strips of rough grassland or even hedgerows provide them with an opportunity to ambush prey.

Hen Harriers are occasionally seen 'hedgerow hopping', a technique adopted often by the more common Sparrowhawk. A hedgerow may offer the only significant cover in otherwise open farmland and so harbour small birds seeking food and shelter. Jinking from one side of the hedgerow to the other during hunting flights increases the chances of taking a bird by surprise.

The Hen Harrier's strategy of roaming widely allows it to take advantage of changes in the quality of hunting habitats during the course of the winter. A recently harvested, weedy stubble field may attract dense flocks of sparrows, finches and buntings to feed on the spilt grain and weed seeds, offering an

attractive place to hunt. No doubt Hen Harriers remember such sites and are likely to return regularly over the coming days. The sites richest in prey may attract two or three birds from the same communal roost site and, exceptionally, as many as five birds have been observed hunting over the same fields. But once the field is ploughed and re-sown with the next crop, the cover will be lost; the seed-eating birds will move on, and so must the harriers.

Danube Delta. For a few minutes the ringtails teamed-up and worked the area together. One flying almost at ground level looking to flush out any potential prey, the other holding back, slightly higher, ready to pounce. It's unlikely these birds would share food but hunting together may help to spook more birds and so improve the chances of at least one of them making a kill. Either way, the tactic was fruitless and they eventually returned to their solo means of hunting.

FLUSHING SNIPE ON THE MEADOW
T.HAVEN

HISTORY AND STATUS IN BRITAIN

The long decline

THERE ARE VERY FEW RECORDS of Hen Harriers from early history, though bones have been found from as far back as the Iron Age, as well as from a few Saxon and medieval sites in Britain. Nevertheless, it is clear that the activities of humans would, initially, have benefited a bird requiring open country. Significant parts of the uplands where harriers now breed would once have supported dense woodland, offering fewer opportunities for the Hen Harrier. And in the lowlands, before humans started to make an impact, large expanses of open country were probably restricted to coastal areas. As more and more woodland was cleared by early settlers, there would have been increasing amounts of open land over which the birds could hunt. Grazing livestock helped keep these areas open and the low intensity of management would no doubt have supported a good prey base for harriers to exploit. It is necessarily speculative but it's not hard to imagine

Quartering the ancient fishponds at Keyhaven, Hampshire.

the Hen Harrier as a common and widespread bird before less positive human impacts began to take their toll.

Given its current status as a bird of the upland moors, it is easy to forget that the Hen Harrier was once also widespread in the lowlands. In the early part of the nineteenth century it was said to breed regularly on Exmoor in Devon/Somerset and in the New Forest heathlands in Hampshire, and there were breeding records from many other counties in the southern half of England, including Kent, Surrey and Sussex. Confusion with the similar Montagu's Harrier obscures the true picture; they were not fully recognised as separate species until 1802. Yet there is little doubt that the Hen Harrier was once widely distributed in central and southern counties as well as in the uplands of Scotland, Wales and the north of England.

Red-legged Partridges keeping a low profile.

The familiarity of the bird meant that it had various common names in different parts of Britain that have long faded from use. The striking plumage of the male is the inspiration for Blue Kite (Scotland), Grey Buzzard (Hampshire), White Hawk and Seagull Hawk (Ireland), and Miller (Cambridgeshire; because of the dusty whiteness that comes from grinding flour). Its choice of habitat is alluded to by Furze Kite (Devon; 'furze' meaning gorse), Gorse Harrier (Sussex) and Moor Hawk.

Birds of prey in Britain have long suffered at the hands of humans, both as a result of changes to their habitats and through the direct effects of persecution. The Hen Harrier is no exception and its fortunes are closely tied to the way it is treated by people. In southern England the species declined initially as its habitat was reduced and fragmented. Wetland breeding sites were lost and large open field systems were subject to 'enclosure', being divided up into much smaller fields, separated by hedges. Huge areas of farmland were treated in this way, and while there were benefits for some hedgerow-loving birds, those requiring large expanses of open country lost out.

Ultimately, it was persecution that sealed the fate of the lowland Hen Harriers, partly by householders concerned about the safety of hens straying from their coop, but also by those responsible for rearing gamebirds for shooting. Anything seen as a threat to partridges or introduced Pheasants was simply not tolerated. Small armies of gamekeepers patrolled the countryside, shotgun in hand, and many birds of prey and predatory mammals suffered dramatic declines. The Hen Harrier is more vulnerable to persecution than most raptors because of its behaviour. It hugs the contours of the land and flies at a leisurely speed, making it an easy target for the shotgun. And it nests on the ground, so that its eggs and chicks are vulnerable to those who have no wish to see young birds take to the skies.

Adult Hen Harriers often try, desperately, to defend their nests by dive-bombing intruders that stray too close. This behaviour has evolved to help prevent predation by ground predators and it is often successful. But against gamekeepers waiting close by with their guns, the results are all too predictable. One particularly merciless technique employed by gamekeepers exploited the parental devotion of the adults. If a nest was found, all the young except one would be killed. The remaining chick was then tethered to a stake to prevent it from fledging. The adults would keep returning with food, allowing ample opportunity to dispatch them. By the middle of the nineteenth century the Hen Harrier had become rare or was entirely absent from most of its former haunts in the lowlands. Even in winter the bird is vulnerable. It congregates in numbers at the same communal roost sites year after year, allowing it to be targeted with relative ease.

Things were little better in the uplands where the shooting of Red Grouse was the primary sporting interest. The railways opened up previously remote areas in northern England and the Scottish Highlands to visitors from the big cities, helping to increase the popularity of the sport. Here, too, gamekeepers took their duties seriously and the Hen Harrier was especially disliked for its predation of grouse and the way it could disrupt shoots by spooking birds as it quartered low, back and forth, across the moors. No fewer than 68 harriers were killed on the Glengarry estate in Inverness-shire in 1837–40. Trumping that is a claim that 351 harriers were dispatched on a grouse moor in Ayrshire, southern Scotland, in just four years from 1850. It may be an exaggeration but, even so, it provides

A rare New Forest snowscape. A ringtail hunts over the few exposed areas of heath, the Beaulieu Heath Bronze Age barrow in the distance.

an indication of the scale of persecution taking place. The Victorian passion for collecting birds and their eggs was a further pressure, increasing as the bird's rarity made it ever more desirable, and encouraging a pursuit of the Hen Harrier into its few remaining refuges. Nests were plundered and the adults either shot close to the nest, entangled in snares or caught in small traps placed nearby. By the early 1900s the beleaguered Hen Harrier survived as a breeding bird in Britain only on a few islands off the west coast of Scotland and on Orkney in the far north. Here, where persecution was less intensive, it clung precariously to survival and bided its time.

Seeds of recovery?

Perhaps surprisingly, it was the Second World War that proved to be a turning point in the Hen Harrier's fortunes. Gamekeepers were called up to fight for their country, requiring them to deploy their firearms skills with a new purpose, and offering some respite for birds of prey. The Hen Harrier grabbed its chance, recolonising the Scottish mainland in 1939 and consolidating its position over the coming decades. Wales and northern England were recolonised over the next 20–30 years, as was the Isle of Man in the 1970s, followed by a rapid population increase there.

Orkney is a vital stronghold for the Hen Harrier. There was concern following a substantial decline towards the end of the last century but numbers have recovered well in recent years as the area of rough grassland, rich in Orkney Voles (see June*), has increased. In 2011, a record 120 females bred on these islands.*

The first national survey in 1988/89 estimated that there were 578 pairs in the United Kingdom (including Northern Ireland and the Isle of Man), the majority in Scotland. By the time of the 2004 survey the estimate had increased to 806 pairs. Welcome though this was, the expanding populations have, once again, come into conflict with gamekeepers and the Red Grouse they seek to protect. Recovery has stalled, and in areas where grouse shooting is at its most intensive, numbers have declined. Take northern England, for example, where intensive grouse shooting is the dominant land-use in many of the places where Hen Harriers try to breed. Breeding was thought to have occurred in sixteen 10-km squares in 1968–72. In the early 1990s perhaps 40 pairs could be found, concentrated in the Forest of Bowland in Lancashire and around Geltsdale in the north Pennines, but with pairs widely spread in Cumbria, Derbyshire, Durham, Northumberland, Staffordshire and Yorkshire. And yet, by 1998, breeding was restricted to just nine 10-km squares.

The latest national Hen Harrier survey, in 2016, found just four pairs in England, representing a new low point. Since then, there have been tentative signs of recovery. In 2018, there were nine successful nests in northern England, although almost all were away from intensively managed grouse moors. In 2021 this figure had increased to 24 successful nests, still a tiny population but offering a glimmer of hope for the future.

In Scotland, too, there have been declines in areas dominated by grouse shooting, not fully compensated by increases in other areas away from grouse moors. Even in Orkney, where grouse shooting does not take place, the population underwent a substantial decline, in this case due to a reduction in high-quality, prey-rich habitat. As sheep stocking levels increased there was a reduction in rough, tussocky grassland which provides such an important hunting habitat, particularly for the smaller males. Thankfully, this population has recovered during the past 20 years as the area of rough grassland has increased again. The most recent national survey in 2016 estimated the overall Scottish population at 460 pairs, a substantial decline from the 2004 estimate of 633 pairs. The smaller populations in Northern Ireland, Wales and the Isle of Man have also fallen away since their peak around the time of the 2004 survey. The likely reasons for this are explored in a later chapter (see *Threats and survival*).

Sadly, there has been no sign of a meaningful recovery in the lowlands although persecution levels here, well away from the grouse moors, are now far lower than they were in the past. There have been occasional isolated breeding attempts in recent years, for example in Gloucestershire, on the Lizard peninsula in Cornwall and in Wiltshire, but no sign of consolidation. Conservationists have seen how

Estimates for territorial pairs of Hen Harrier based on national surveys

	1988/89	1998	2004	2010	2016
England	18	19	11	12	4
Wales	27	28	43	57	35
Scotland	479	436	633	505	460
Isle of Man	44	49	57	29	30
Northern Ireland	10	38	63	59	46
Total	578	570	806	662	575

Figures from Sim *et al.* 2001, 2007; Hayhow *et al.* 2013; and Wotton *et al.* 2018.

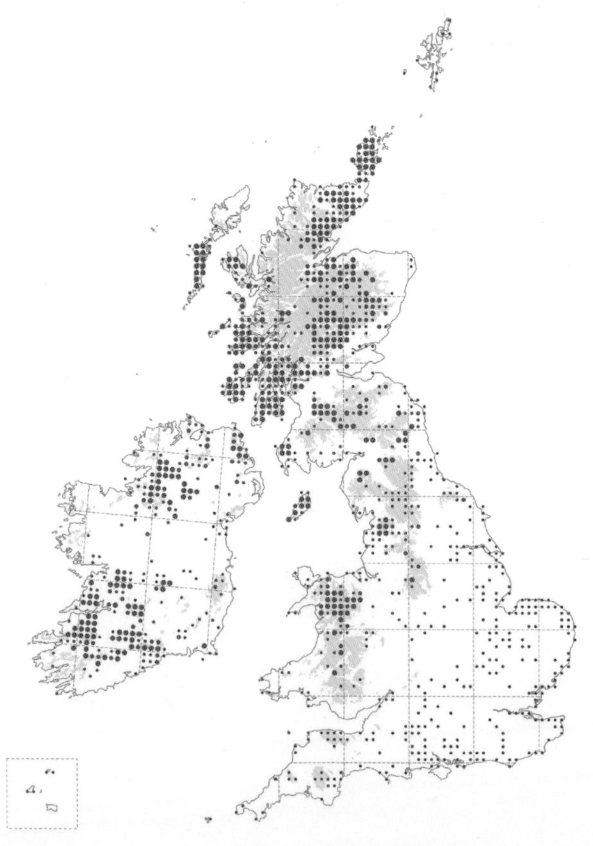

The Hen Harrier is restricted to upland areas in the breeding season, with a patchy, fragmented distribution away from its strongholds in western Scotland and the Scottish Highlands. The red dots indicate breeding, with small grey dots showing sightings of transient birds in summer away from breeding sites. (Map from Bird Atlas 2007–11, *a joint project between BTO, BirdWatch Ireland and the Scottish Ornithologists' Club – reproduced with permission from the British Trust for Ornithology.)*

successful Hen Harriers can be in lowland areas dominated by farmland on the Continent, where they sometimes nest in crop fields, and have been exploring the options for a reintroduction into southern England. It is a contentious idea. Many people believe that the focus should be on tackling illegal persecution and allowing the bird to increase naturally, rather than resorting to such a dramatic intervention.

There is less information about historical changes in numbers of wintering birds, but it is clear that the Hen Harrier has long been a widespread, if rather uncommon, bird in Britain during the winter months. While small numbers of birds remain in their upland breeding areas, there is a general shift to lower ground, and coastal sites are often favoured. East Anglia and south-east England are winter strongholds. England alone probably supports several hundred birds in most years. In the early months of 1979, when hard weather encouraged birds to move south and west to more benign conditions, it is thought that around 750 birds were present in England. An apparent growth in the wintering population during recent decades has coincided with increases in the breeding population on the near continent, especially in the Netherlands, which acts as a source of migrants, boosting numbers of wintering birds derived from the British breeding population.

MARCH

MARCH IS A MONTH OF TRANSITION for the Hen Harrier. Winter finally draws to a close and attention turns, tentatively at first, towards the breeding season once again. The communal winter roosts break up and wintering areas in the lowlands are abandoned as birds begin to drift back north to their breeding sites. In parts of Britain, the transition is hard to detect as individuals may be encountered close to their breeding areas in any month of the year. The larger females, especially, are able to feed on bigger prey such as grouse that can tough it out for the winter in the harsh conditions found in the uplands. In contrast, across much of northern Europe, birds are returning from wintering grounds hundreds, even thousands, of kilometres to the south. Upland landscapes that have seen few birds of any kind during the winter gradually regain their songbirds and waders, together with the predators that come to feed upon them.

Breeding habitats

Based on recent history in Britain, we tend to associate breeding Hen Harriers with expansive upland moors, particularly of an altitude between 150 and 600 metres (exceptionally up to 700 metres), where patches of tall vegetation provide suitable places in which to hide a nest. Some of the best areas are protected through designation as Sites of Special Scientific Interest (SSSIs), and there are also Special Protection Areas (SPAs) for Hen Harriers, classified under the EU Birds Directive, with a status that is now less secure following Brexit. Sadly, the SPAs in England and southern Scotland have lost most of their breeding harriers due to persecution, despite their protected status.

Mature heather is often used for breeding and provides cover to help conceal the nest and a degree of shelter. Moors on which the older heather has been lost due to overgrazing by livestock or excessive burning become less suitable as breeding sites. Taking a historical view, and looking beyond British shores, it is clear that the Hen Harrier is a flexible and adaptable bird and will take advantage of a wide range of different habitats. In the past it bred widely in Britain in the lowlands, making use of rough vegetation in wetlands, heathlands or even scrubby grassland. Further afield, in the Netherlands, it breeds in grassland

The Hen Harrier is an adaptable bird and will breed in a wide variety of habitats provided its key requirements are met. It needs open landscapes rich in food, sufficient cover in which to site its nest and to help mask its approach when hunting, and sites that are not subject to excessive disturbance by humans or high densities of nest predators. In Britain, it is primarily a bird of the uplands, though steep, mountainous terrain is unsuitable for hunting and so is avoided.

habitats within sand dunes as well as the edges of wetlands. And in parts of Europe, including eastern France and northern Spain, it has adapted to take advantage of the dense vegetation and open spaces provided by arable farmland. Here it breeds in fields of crops, usually winter cereals, often in the same areas as its smaller relative, the Montagu's Harrier. Both species are able to rear young by hunting prey in the surrounding crops and field margins. Studies show that although there is much overlap in the diet, the Hen Harrier tends to take larger prey and fewer invertebrates than its smaller relative.

Crop nesting birds are vulnerable to farming operations, particularly harvesting, but Hen Harriers nest earlier than Montagu's Harriers and so their young are more likely to have fledged by the time the crops are gathered in. For late nesting pairs, intervention by raptor workers who rescue the young or mark the nest area so it is left uncut, offers the only real hope of surviving the harvest. Occasionally, an isolated pair has bred in arable farmland in southern England, hinting at the largely untapped potential of this habitat here in Britain.

Hen Harriers have adapted to nest within blocks of conifers. Before the trees grow too tall, they nest in the rough grass anywhere within the plantation. Later, they favour clearings or wide rides through the trees, or areas where the trees have been cleared and replanted. Nests in forests are hard to find because the adults often fly along rides and tracks below the tops of the trees to reach the nest site. They are also less prone to casual disturbance and persecution as few people venture into these dense, forbidding forests.

Hen Harriers (left) and the smaller Montagu's Harrier (right) breed in the same areas in parts of France and northern Spain, making use of the artificial vegetation provided by arable crops.

Once the clag set in, we were reduced to glimpses of activity. This bird just visible as it made its way along the edge of the plantation. Near to the nest.

The Hen Harrier's adaptability is evident in the uplands of Britain. Over the last century or more, large areas of moorland have been planted with non-native conifers. Dark blocks of trees now scar the landscape, their angular boundaries making it look as if they have been painted onto the hillsides when seen from a distance. But even here, the Hen Harrier is able to take advantage, at least in the early stages while the young trees are growing. Thick grassland often grows up between the trees, particularly where they are fenced to keep the deer out. This supports good populations of small mammals, as well as providing a secure, often disturbance-free, site for the nest.

Plantations have provided something of a refuge for the Hen Harrier, away from the gamekeepers who patrol the surrounding grouse moors. Recently, their importance has declined somewhat, as there has been less new planting, and replanted forests, following felling, tend to have less rough grassland and so support fewer small mammals. But in western Scotland, Northern Ireland and the Isle of Man a significant proportion of the population breeds in wooded habitats of some kind, including conifer plantations, naturally regenerating scrub and woodland, as well as the increasingly large areas where native trees have been replanted.

Setting up territory

There is limited evidence from countries where the breeding areas are abandoned completely (and so the patterns of arrival are clearer) that the male Hen Harriers tend to return a little in advance of the females, as happens with many territorial species. They risk losing out if they are late back and all the best sites have already been occupied. The females can afford a slightly more laidback approach, biding their time before seeking out a suitable territory-holding male to join.

Established pairs sometimes return to the territory they used in the previous year, although re-sightings of marked birds have revealed that they will also move long distances to a new breeding area. One female, identifiable from the wing-tags fitted by scientists (see below), moved 188 kilometres between breeding attempts in Scotland. It is thought that birds are more likely to move if they have lost their partner or if breeding was unsuccessful in the previous summer. It is unlikely that many pairs stay together from one year to the next, though good information on this is lacking.

A female drifts across to meet a male and both perform a beautiful aerial act of courtship – an intricate weaving dance.

A male pushing into a strong wind. Donald Watson, who studied and painted Hen Harriers over many decades, memorably described birds behaving in this way as 'like tiny arrow heads' when viewed from a distance, neatly capturing the shape formed by their long tails and flexed wings.

This early in the breeding season, the activities of Hen Harriers are heavily dependent on the weather. In good conditions they spend time flying around potential breeding sites from as early as February, though March is more typical. But if harsh conditions return, they may retreat temporarily back down to lower ground. Aerial activity at this time tends to be rather unspectacular. It may consist of little more than a bird, or an established pair, circling slowly over the breeding area as if contemplating their next move. They are capable of reaching a great height on these flights, leaving observers straining to follow them as they become mere specks in the sky, or even disappearing into the clouds.

Sometimes a pair will fly very close together as they circle, the male usually slightly above the female, and there may be some playfulness, one bird dipping towards the other, wing-tips almost touching, or rolling to one side to present its talons. These high, circling flights offer a contrast to the usual purposeful hunting flights low over the ground. They help birds already paired to maintain a strong bond, and they no doubt also help them become more familiar with their breeding territory and any other harriers that might be in the area.

First-time breeders

As with many birds of prey, there is a tendency for first-time breeders to settle reasonably close to the site where they were reared, though this is by no means a hard and fast rule. The Hen Harrier shows more flexibility in this regard than some large raptors, and individuals with a more adventurous streak may settle hundreds of kilometres away from their birth-site. A study in Scotland involved fitting wing-tags to nestlings and then recording the location of their first breeding attempt. There was much variation but both sexes moved over 10 kilometres on average, with a tendency for males to settle further from their original nest site than females.

Evidence from wing-tagged birds shows that many females attempt to breed when they are only one year old, whereas males usually wait until they are two. This reflects the fact that males are responsible for finding much of the food for nestlings and so are more likely to be successful when they have an extra year of experience under their belts. But if there is a territory and a willing female available then sometimes even the youngest males will give it a go. Even if unsuccessful, they at least gain valuable experience that will benefit them in the years ahead.

It is likely that first-time breeders mainly select the habitat where they were reared, even if they move to a new area to breed. A bird fledged in a cereal crop learns that this is a suitable place to nest and looks for similar habitat when it comes to breed. Equally, a bird reared on heather moorland will probably return to the uplands to seek out a similar landscape.

Male birds only occasionally breed in their first year and when they do so they are less likely to be successful. These young males have mainly brown plumage early in the season and are difficult to separate from the females. But, as moult progresses, they begin to gain a few of the grey feathers that are characteristic of the adult males (lower bird).

APRIL

THE PACE OF LIFE PICKS UP in early April as the time fast approaches for birds to commit to a nest site and begin the serious business of raising the next generation. Aerial activity becomes more frequent and purposeful as birds signal ownership of their territories. Those that are still unpaired have one final chance to seek out a potential mate. If weather conditions are good, this is the best month for fieldworkers to spend time in the hills looking for harriers and trying to pinpoint the sites where they will breed. Nest-building often begins in April and the earliest pairs will already have a clutch of eggs by the end of the month.

Food in early spring

One of the most important considerations when choosing a place to nest is the food supply. It's one thing for an individual bird to catch enough prey to keep itself alive. But later in the season, if all goes well, the adults may have four or five rapidly growing young to support. When the chicks are small the female spends most of her time at the nest, and it falls to the male, almost single-handedly, to find enough food for the whole family. This will only be possible in an area with a rich supply of prey.

In the majority of the upland breeding areas in England and Scotland, Hen Harriers rely on two key prey species when selecting a nest site. If an area supports high densities of Meadow Pipits and Field Voles then harriers are drawn to it, reassured that food is plentiful. To complicate matters, the Field Vole is prone to large, cyclical fluctuations in numbers, so populations vary greatly from year to year. A site with a rich food supply in one year may be far less suitable a year later, so harriers must be flexible in their choices.

If the densities of these key prey species are patchy then harriers will be attracted to the places where they are most abundant. This leads to aggregations of nesting birds, with hotspots where a number of pairs try to breed and other places where harriers are scarce or absent entirely. This may give the impression that the Hen Harrier is semi-colonial, but it is probably the appeal of a good food supply that is the driving force behind these clusters of birds, rather than harriers necessarily

The Field Vole and Meadow Pipit (left) are key prey for the Hen Harrier across much of Britain. They occur at high densities in places where there is a good mix of rough grassland and heather, favouring sites where the vegetation is not too heavily grazed. The slightly larger Skylark (right) is often common in the same areas and is also a frequently taken prey species.

wanting to be close to others of their own kind. Polygynous breeding, where a male is paired to more than one female, also leads to clusters of nests within a small area (see *May*).

The skydancing display

The Hen Harrier has long been admired for its spectacular display flights. The term 'skydancing' was first coined back in the 1960s by the late Frances Hamerstrom (a graduate student of Aldo Leopold, the famous American ecologist and environmentalist) who was studying the closely related Marsh Hawk in the United States. It captures perfectly the exuberance and sense of freedom that seems to possess the birds in early spring as they take to the skies.

The display begins with a bird circling up to a considerable height. It then uses deep 'rowing' wingbeats to pick up speed before making the first dip vertically down towards the ground. As the display progresses, the dives become longer and the wings thresh more wildly, until it seems that the bird must surely crash into the ground. But, at the last second, it performs a dramatic 180 degree turn and swoops back, vertically upwards, in readiness for the next instalment. This skydancing display is mostly associated with males, and their pale plumage helps them to stand out against the backdrop of dark, heather-clad hills, even on a dull day.

The contrast between the grey back, jet-black wing-tips and the much paler underparts appears exaggerated in the rapid, swooping changes of direction at the top and bottom of each dive. Some watchers are reminded of a Lapwing's display, both in the contrast between the darker upperparts and pale plumage below, and in the frenetic rising and falling flights above the ground.

Female harriers also indulge in skydancing, although this is seen less often and tends not to be so prolonged as in males. It probably serves a similar purpose. Females are equally keen to impress members of the opposite sex in order to attract a new mate, or hold on to an existing one. They are probably also sending out a message to rival females who may think of competing for 'their' male's attention. As discussed in the following chapter (*May*), this is a perennial concern for a female Hen Harrier because of the polygynous behaviour of some males. A female would rather have a male all to herself, so that he can focus his food provisioning solely on her and her young, rather than having to share. If she fails in that aim, she will at least want to be the primary female to which the male will devote the majority of his attention.

When the biological reasons behind these displays are understood, it becomes easier to appreciate why they can be so prolonged and spectacular. Such vigorous displays require considerable effort and they therefore provide a true indication of the fitness and quality of the bird involved. A half-hearted effort is unlikely to impress potential mates or rivals, and participation in the breeding season ahead will only be assured with a good performance. Little wonder then, that the birds throw themselves into it with all the energy and enthusiasm they can muster.

The Hen Harrier's skydancing display may seem like an expression of pure joy when viewed from a human perspective. Yet it is carried out with a very specific purpose in mind, in order to hold on to a breeding territory and appeal to other birds. Unmated males have been recorded making over 100 consecutive dives during a bout of skydancing, as they desperately try to attract the attention of a passing female. In the closely related Marsh Hawk, studies have found that the males displaying most often and most vigorously tended to attract more females. If there is joy and emotion involved it is primarily in the people lucky enough to be in the right place at the right time to witness the spectacle. It is a sight that lingers long in the memory.

Skydancing is all about movement, speed and agility – 'like a bird possessed' as the renowned Hen Harrier artist Donald Watson put it. At the top of each vertical ascent the bird may flip over onto its back momentarily (in a kind of reverse loop the loop) before beginning the next downward plunge. Jet-black wing-tips contrast with the rest of the plumage, and loose, exaggerated wingbeats draw the eyes of the observer as the bird flings itself recklessly towards the ground.

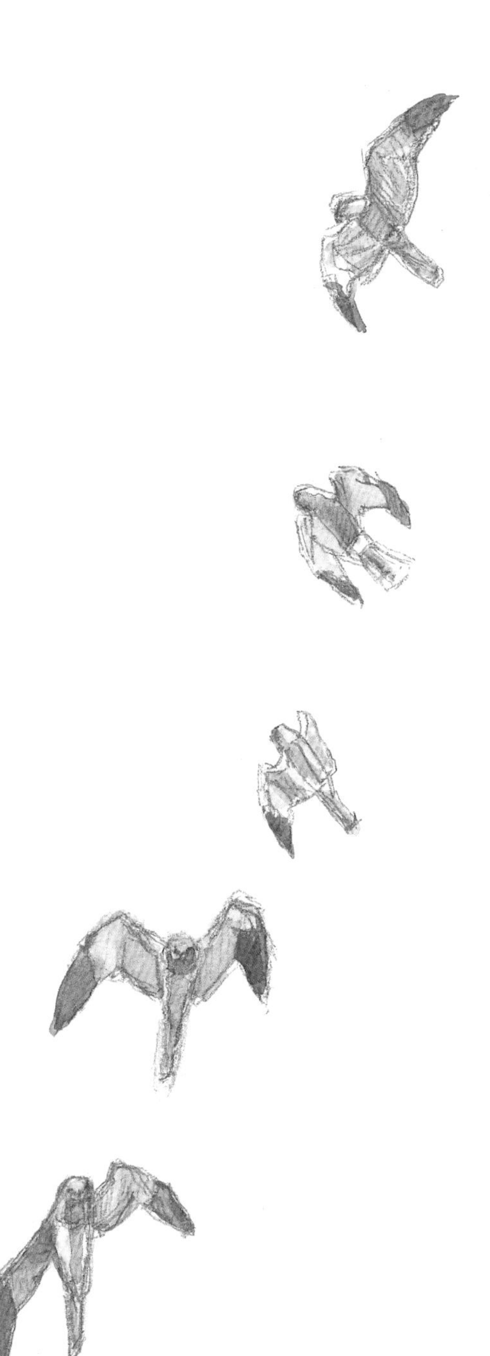

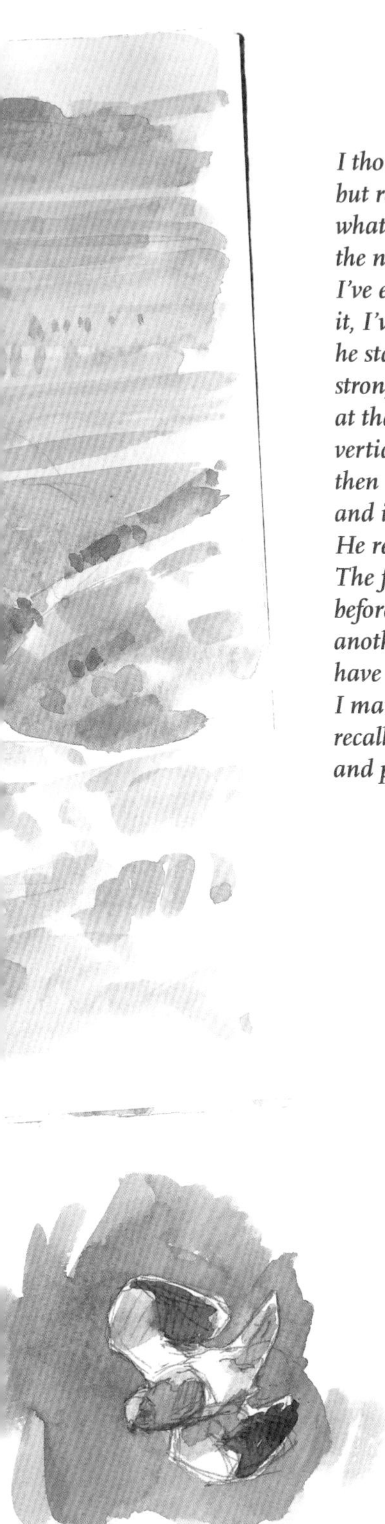

I thought we were too late in the season, disappointed, yes, but resigned to not seeing it happen. Then out of nowhere what seemed to be a routine food pass and escort back to the nest quickly turned into the most stunning skydance I've ever had the pleasure of witnessing (you've guessed it, I've only seen one, but boy was it good). To warm up he started to 'roller coaster' beside her, before flying with strong, stiff wingbeats that gave him height. Twisting at the peak of his climb, head held level, he plunged vertically at breakneck speed towards the ground, and then with astonishing agility he pulled out of the descent and into a vertical ascent as if tugged at by a puppeteer. He repeated this several times. Simply breathtaking. The female carried on nonchalantly – seen it all before, he's just showing off. It was then that I noticed another harrier in the far distance. Could its presence have triggered this display as a territorial warning? I manically scribbled down as many shapes as I could recall and probably some that I hadn't, such was the rush and privilege of catching nature at its most spectacular.

Vocalisations

Away from the breeding grounds, the Hen Harrier is not a vocal species but this changes when the male and female interact close to their nest site. Displays are usually accompanied by bouts of chattering or 'yikkering' calls, the female having a higher pitch than the male. When courtship is intense the female may give a distinctive, high-pitched call variously described as a wail or whistle, not dissimilar to a Wigeon or the mewing call of a Buzzard. This appears to be a sign that she is ready to mate and she may even pursue the male, calling with increasing intensity and persistence if her advances are rebuffed. The same call is also used when the female is about to receive food from the male. One of the calls made by male birds has been described, very aptly, as a 'chuckle'. It can sometimes be heard as the male brings in food and is rather understated against the backdrop of the higher-pitched, more insistent, soliciting calls of the female.

Copulation and nest-building

Early in the season, copulation often follows the delivery of a food item by the male to the female. Once the food is exchanged (see *June*), the birds land at the nest or close to where it will subsequently be built and the female often calls to indicate that she is willing to mate. No doubt the close association between food delivery and mating rights encourages the male to hurry back to his hunting duties at the earliest opportunity.

Both sexes have a role in deciding where the nest will be built, but it is the female who does most of the work and presumably she has the final say. In parts of the range, including Orkney, nests are often in damp areas and are built-up higher than those on drier ground; these waterlogged sites may provide some protection from ground predators. As suggested by its alternative name, this habit is more commonly adopted by the Northern Harrier (or Marsh Hawk) in North America. In Europe it may be competition from the Marsh Harrier, a

Notebook entry: Stick gate – male comes in carrying long heather stick in bill. Drops it in nest. Female removes stick, takes it away to earthy mound, some 100 metres from nest. Brings stick back to nest. Male goes to nest, takes stick back to earthy mound. Female grabs stick, returns it to the nest. Male goes to the nest, lands in nest and is then turfed out of nest. Returns to earthy mound and starts to tear at the ground. Clearly this is the behaviour of an inexperienced male learning the nesting trade, but in my head I hear the female: 'back-off sunshine this is my job, go and do yours. Voles will do nicely.'

bird closely associated with wetland habitats, that has resulted in Hen Harriers tending to select drier sites across much of the range.

Nest material is gathered from within about 200 metres of the nest and may be carried in the beak or the feet. Larger items can be very obvious, trailing behind a bird in flight, and giving a clear indication that breeding is under way. While most of the burden of nest construction usually falls to the female, the male at least helps to gather material. Lone birds will sometimes build a complete nest unaided. These have been referred to as 'cock' nests when built by an unpaired male and may serve as an added enticement to help attract a potential mate. The nest is usually tucked away securely within a patch of tall vegetation, though when viewed close up it is often in a small open spot, facilitating access for the adults flying in and out of the site, and giving the young room to move about later in the season.

The nature of the site chosen varies depending on the habitat available, but in the uplands, patches of well-grown heather are often selected. Where this is lacking, Bracken, rough grass, rushes or even gaps between small trees or scrubby bushes may be used. At these sites, the vegetation grows up considerably during the course of the breeding season and the nest becomes more effectively hidden as the weeks go by. Gently sloping ground is often chosen which may help ensure the site does not become waterlogged. The birds are highly flexible in the material used to build the nest, making use of Bracken, plant stems, heather and twigs from both broadleaved and coniferous trees, depending on availability.

Heather is often the major component of nests in most of the breeding range in Britain and is the obvious choice of material to help hide a nest within patches of the same plant. Nest construction is rather loose, but the nest looks neater once soft material has been added to the centre to provide a lining for the eggs. If Bracken or dead grass is used, the central lining stands out from the darker twigs used for the surrounding structure. Nests are not reused from year to year but a new nest is sometimes built in the same patch of vegetation in the following year, and good sites may be occupied regularly over many years.

Nesting in trees

A remarkable and unexpected discovery was made by raptor workers in Antrim, Northern Ireland in the early 1990s. Hen Harrier nest sites were found up to 13 metres from the ground in conifers. This has been a regular occurrence since first discovered but has, so far, remained unique to this area. Such nests have a major advantage in one respect: they are far less accessible to ground predators. But there are disadvantages, too. The rather flimsy nests are more exposed in poor weather, increasing the risk that eggs or young will become chilled. And nestlings are vulnerable to falling to the ground where they are unlikely to survive unless they happen to be found and rescued by fieldworkers. The young lack a strong instinct to avoid straying from the nest platform, which is hardly surprising in a bird adapted for ground-nesting. Overall, tree-nesting birds in Northern Ireland tend to rear fewer young than ground-nesters and, perhaps as a result, the trait has declined in recent years. That it happens at all is probably a reflection of the degraded and overgrazed condition of the surrounding moorlands, resulting in a lack of typical nest sites and forcing birds to think beyond their usual comfort zone.

The Merlin, like the Hen Harrier, is normally a ground-nesting bird in Britain but now regularly adopts old corvid nests in conifer plantations close to its favoured moorland haunts. In contrast, tree-nesting harriers have been found only in Antrim, Northern Ireland. Nests are constructed on the flattened tops of poorly growing conifers. They are rather flimsy, in typical Hen Harrier style, and are at risk of collapsing and crashing to the ground in wet and windy weather. Tree-nesting Merlin (left) and Hen Harrier (middle and right).

MAY

IT FINALLY STARTS TO FEEL LIKE SUMMER as April gives way to May. The days are longer, and while the cool, showery conditions prevalent in April can return at any time, the sun is now higher in the sky and better able to exert its power when the clouds disperse. In Britain, the majority of Hen Harrier clutches are laid in early May, though there is much variation depending on the local food supply, prevailing weather conditions, and the idiosyncrasies of individual birds. By the end of the month the longest day is just three weeks away and pairs that lay early, in April, will already have small chicks – the long and difficult task of rearing the next generation of young harriers now begins in earnest.

Incubation and hatching

Normally, a clutch comprises between three and seven eggs (with an average of about five) and they are laid at intervals of between one and three days. Larger clutches of up to 12 eggs have been recorded but probably involve contributions from more than one female or situations where a replacement clutch has been laid in the same nest as older, unhatched eggs. The adult female is responsible for incubating the eggs, and her dull brown plumage helps keep her hidden away among the vegetation. Each egg requires about 30 days before it hatches but because incubation begins well before the final egg is laid, the chicks emerge at intervals, the first gaining a considerable advantage over the others. For a large clutch it may take more than a week for all the eggs to hatch and, as a result, there will be a big size difference between the first and last chicks in the brood.

This is a sensitive period in the Hen Harrier's lifecycle, when many nests fail. The length of time required to incubate eggs and rear young means that there is only time for one breeding attempt in each summer, so the stakes are high. Only

The brown plumage of the female keeps her well hidden during the long weeks spent at the nest. The male is much paler but spends little time at the nest site. It has been suggested that his plumage may also help to conceal him – in this case when viewed by prey from below against a bright sky. The dark wing-tips may help in the deception by breaking up his outline and obscuring the overall shape of the wings.

if the nest fails at an early stage of incubation will it be possible for the female to lay a new clutch and try to salvage something from the season.

Eggs and young chicks are vulnerable to opportunist avian predators including corvids and gulls, and to birds of prey patrolling overhead. Then there are the ground predators such as Foxes, Stoats and even Mink that roam the moors, always on the lookout for nests to plunder. A camera at a nest in Scotland showed a Fox arriving at night, flushing the female but quickly leaving. A Short-eared Owl then landed at the nest but also left empty-clawed, before a third predator, a Long-eared Owl, entered the fray. In the absence of the female harrier it was able to remain at the nest undisturbed as it devoured the small chicks. An attack by an Eagle Owl was caught on camera in northern England. In this case, the female harrier was attacked and, perhaps understandably, failed to return to the nest, causing it to fail. If a Fox manages to sneak up on a female at the nest site then she may be killed, ending, in an instant, any hope of young being reared.

Cool and damp weather also presents a threat, as eggs and young can quickly become chilled, particularly if the adult female has been disturbed and is away from the nest. Small, downy chicks are especially vulnerable as they rely on the female to provide them with shelter, and to keep them warm and dry. The male usually leaves nest-defence and chick-minding duties to the larger female but nest cameras have shown that males do occasionally attend to keep a watchful eye on small chicks while the female is away feeding.

A brood of three small young in the nest, the smallest of which is only a day or so old. They have a covering of pale, pink-washed down which contrasts with the dark eyes and tiny black bill. After about a week, this is replaced with a longer, dingier down. When small, the chicks rely on the adult female to tear food into small pieces which are fed to them beak to beak. Two eggs remain at the back of this nest and may yet hatch. If so, an abundance of food will be needed if all five young are to survive.

Defence of the realm – both adult harriers belligerent and wary. Out of the murk a Goshawk emerges. The male harrier flies above it and cautiously escorts the hawk off the premises; much chittering among the harriers, long after the danger had passed.

A pair of Carrion Crows had chosen to nest in the heather quite close to the harriers. The female was not too happy about her neighbours as a possible threat to her chicks. She frequently flew to a rock that overlooked the crows, giving them a good chittering to. Once flying swiftly from the rock at the sitting bird. Given little chance to defend itself, the crow came off second best from the scrap, minus a good number of feathers, its days probably numbered. Interesting to note the similarity in size between the birds.

Here, another pair of harriers had taken up residence on an adjacent hillside. The male from that pair makes its way slowly, close against the hillside towards the nest. Our male spots him and makes his way over towards him – just checking him out. Then a male from a third pair on the opposite side of the valley also rises above the ridge. Quite something to watch three males in the air at once.

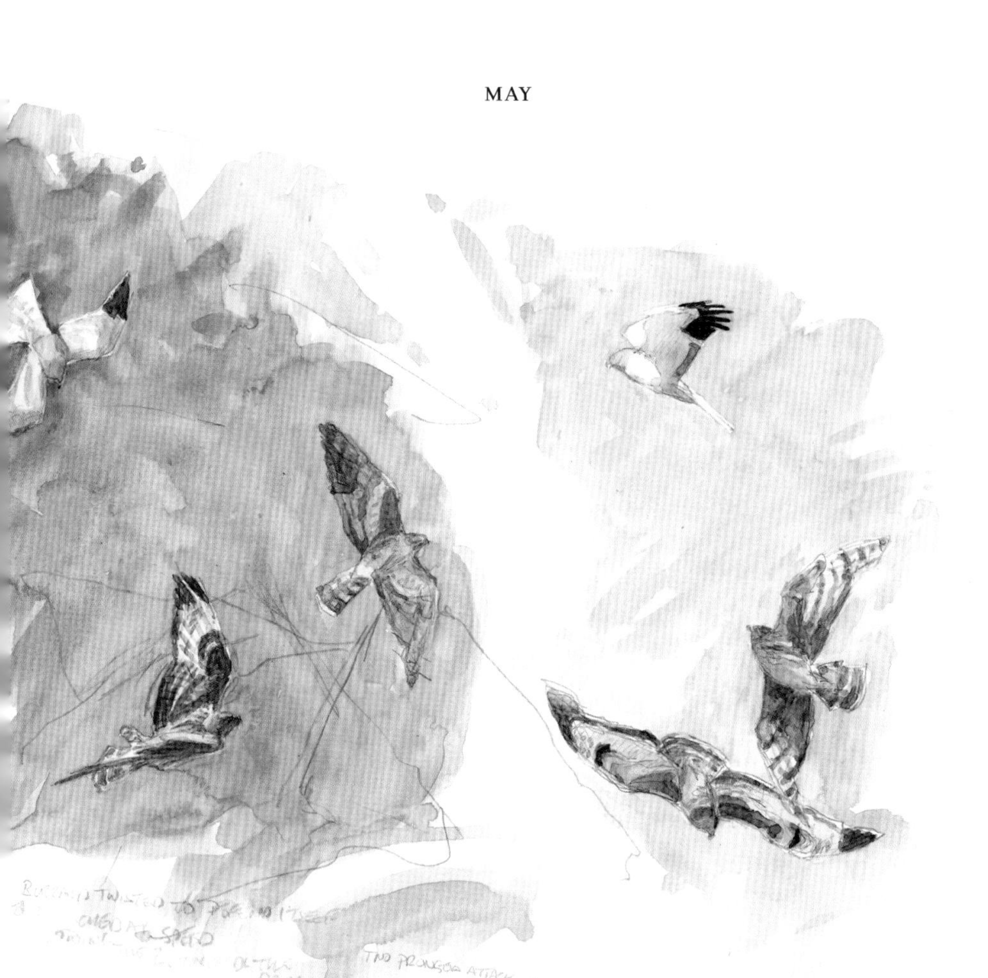

A Buzzard ventures too close to our birds; it's not welcome. The female is onto the situation in an instant – taking sweeping passes at it, making the Buzzard feel uncomfortable, sometimes flicking onto its back, claws raised in defence. The male joins in, forming a two-pronged attack, both hitting the Buzzard from above, driving it towards the ground. Plenty of chittering. The Buzzard breaks free from the bombardment and clears-off pretty damn quickly. The harriers, still not happy, pursue it until it is out of the valley, still chittering away.

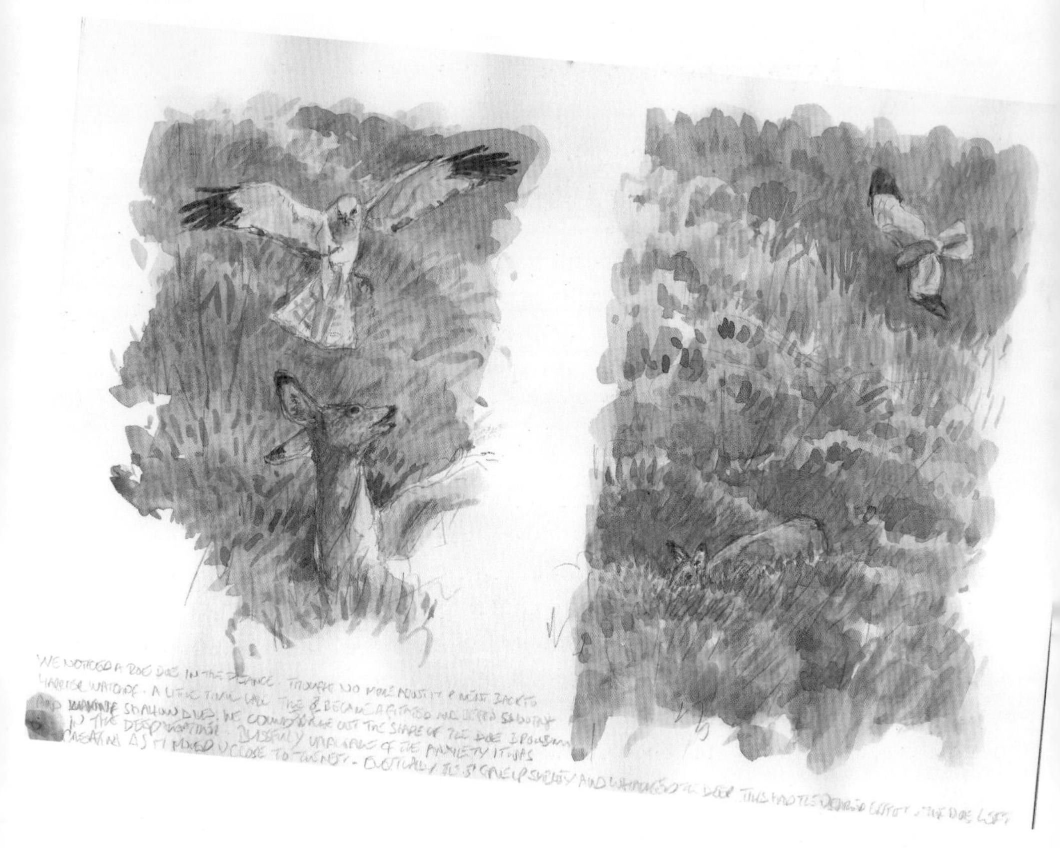

We noticed a Roe Deer doe in the far distance – nice, thought no more about it. A little later the male became agitated and set about circling the nest. He became more and more excited and started swooping at the nest. It was then we picked out the doe again, merrily munching away at the heather, oblivious to the fuss occurring around her. She was getting perilously close to trampling the nest. It all became too much for the male and with a couple of well-aimed flicks of his talons on her ears, the doe changed direction to a more agreeable course. Apart from a 'bemused' look as if to say 'What?' I'm not sure if the deer was at all aware of the stress she had caused. Exiting the scene as she entered, blissfully munching away.

Legal protection and nest visits

The Hen Harrier has been given special protection under the Wildlife & Countryside Act, which makes it an offence to disturb birds when they are at, or near to, an active nest. Breeding pairs vary greatly in how they react to disturbance by humans. Most pairs can adapt to a certain extent if they become used to seeing people whose behaviour is predictable, such as the regular passage of visitors on a well-used footpath or vehicles passing by on a road or forest track. But in more remote sites, or if humans appear unexpectedly, the adults may leave the nest, or fail to return to it to deliver prey, when people are as far as 400 or 500 metres away.

The reaction of adult birds to a human intruder at the nest itself also varies. Fieldworkers visiting a nest under licence to check the contents or ring the young may be mobbed, and occasionally even attacked, by adults (usually the larger female) circling low overhead, chittering frantically in alarm. Eddie Balfour, a renowned fieldworker in Orkney, once lost his trademark beret to an aggressive female which lifted it from his head and deposited it in the heather 50 metres away. The standard field guide for raptor workers (see *Further reading*) recommends that head protection should be worn as a precaution.

This is a behaviour that has evolved to warn off potential predators or large animals that might accidentally trample the nest. It is no doubt effective at making a Fox or even domestic livestock think twice about coming any closer, but it causes real problems in an age when gamekeepers armed with guns are the major threat. The nest contents may be destroyed and the adults themselves will be killed if they are too bold in seeking to protect their young.

Persistent persecution by humans over a long period has probably influenced this behaviour, selecting for greater wariness in the breeding adults. In areas where persecution is intense, the bravest birds will have been weeded out of the population, rarely able to pass on their genes for bravery and aggression to the next generation. In contrast, females that slip quietly away from the nest and leave the area quickly will survive, and may even avoid giving away the nest location. This helps reduce the threat from people, but it will mean that defence of the nest against other predators becomes less effective.

Polygynous breeding

In most large raptors the typical breeding arrangement is a more or less stable pairing between a male and female. Each has their own role to play. Females generally spend more time close to the nest, incubating the eggs and protecting young chicks. Males have greater responsibility for hunting and providing food for the whole family, and spend more time away from the nest. While this familiar pattern, based on a pair of birds, is commonly seen in the Hen Harrier, there are occasions when things get just a little more complicated.

A male bird, as in several other harrier species, may be polygynous and attempt to breed, simultaneously, with several different females. In Orkney, Eddie Balfour's long-term study showed that there were more adult females than males in the local population and rates of polygyny were consequently high. In a large sample of nests, no fewer than 88 per cent involved polygynous males. One male was found breeding with six different females, though 2–5 was far more common.

The nests in polygynous groupings are usually clustered together with the closest just a few hundred metres apart, but as much as 3 kilometres or more may separate the outliers in the group. In the Netherlands, a male that was identifiable by damage to a wing feather was seen provisioning no fewer than seven females in an area with a very rich food supply. One failed at an early stage but the other six managed to rear 18 young in total. It is impossible to know if all were the offspring of the same male bird but this is, nevertheless, a rather impressive output for one breeding season.

Polygyny is an interesting breeding system because there is an inherent tension between the males and females. If a female harrier has found herself a high-quality male, the last thing she wants is to have to share him with other females. Indeed, females are often seen aggressively chasing other females away from their breeding area. In contrast, a high-quality male has much to gain from the arrangement as it gives him a chance to contribute many more young to the next generation. If food is plentiful then all the females in his harem may be able to rear young, even if his efforts in providing food are spread rather thinly. If food is short, he will focus his hunting efforts towards provisioning one partner, referred to as the 'alpha' female, so at least one brood has a good chance of fledging successfully. In effect, he is hedging his bets and can adjust his behaviour according to the available food supply.

The situation appears far less rosy from the female perspective, especially for those that are further down the pecking order for the male's affections. Such birds are likely to receive less support through the provision of food for themselves and their young, as well as less help in defending the nest site should it be threatened by predators. This is a major problem in a species whose young are confined to the nest and vulnerable to chilling and predation when small. Somehow, these females must look after and protect their broods as well as find sufficient food for them with little or no help. They may try to focus on larger prey to minimise the time they are away from the nest, while still bringing in enough food. But it is no easy task.

Why, then, do females put themselves in such a difficult position? After all, the choice of who to mate with rests with the female bird as much as it does with the male. The reality is that they are seeking to make the best of the situation even if it is not ideal. If high-quality males are in limited supply, pairing with one that has other partners is a better option than the alternatives. It is certainly better than missing a breeding season completely with no chance of rearing young. And it may also be better than pairing with a poor quality or inexperienced male, even if she would have him all to herself. Her young will inherit genes from a lower quality male, and he may be no better at provisioning her with food than a high-quality male with several females to satisfy.

Every female no doubt endeavours to become the alpha bird in the arrangement. But if they fail in this, they have no choice but to make the best of it and try to rear as many young as they can, despite being disadvantaged. It should not be forgotten that males, too, can lose out in this mating system. If the best males monopolise the females by breeding with several simultaneously, the lesser males may be left struggling to breed at all. There are winners and losers for both sexes.

Polygyny as a mating system is influenced by the open landscapes used by harriers. In habitats with more cover and obscured sightlines it would be more difficult for birds to keep a close eye on each other. Males would struggle to keep track of several females while trying to make sure that they hadn't mated with rival birds. A male would not want to waste time helping to rear young that might not be his own. The complexity of the Hen Harrier's breeding system can make life difficult for fieldworkers trying to work out how many birds are nesting in an area. They may pinpoint a breeding area based on displays by a male bird early in the season and later find an active nest site close by. But is it the only nest or does the male have other females that may be nesting anything from 200 metres to as much as 1 kilometre away?

Overleaf: The male floats by a stone sheep pen. The female joins him and for a moment they dance together – beauty and elegance lifting a dull, grey day.

25/6/19 - 11.20am

♀ IN COMPANY BEFORE DRIFTING

View from the hillside. Sketchbook snippets. RP.

Northern Marsh- and Heath Spotted-orchid. In some places the Northern Marsh-orchids lined the trackways, forming a guard of honour.

A woolly bear caterpillar (Garden Tiger moth) crosses the track.

We didn't note many butterfly species on the moors, but those we did see appeared in good numbers, including Green Hairstreak which was the most numerous species noted on the moors we visited. Mainly tucked away out of the wind in Bilberry.

Ling (left), also known as Common Heather, dominates our upland heaths. Cross-leaved Heath is found mainly in the damper places, with Bell Heather (right) in drier areas among the Ling.

Beautiful Emperor Moths were on the wing. Brightly coloured males belting low across the Ling trying to catch the scent of a female. Seldom landing.

Offering more opportunity to study than the Emperors, were the numerous Northern Eggar caterpillars munching among the heather.

Green Tiger Beetles scurried ahead of our walking boots.

Getting to the high moors often involved long old walks. None was ever short on interest though. A female Wheatear greeted us with a 'chac' and kept us company, flitting and leading us along a section of drystone wall. A friendly companion. In reality she couldn't divert us away from her nest fast enough.

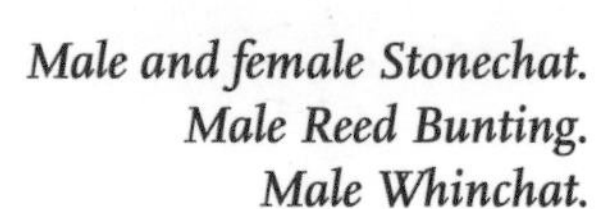

Male and female Stonechat.
Male Reed Bunting.
Male Whinchat.

Trying to avoid us as it used the meandering brook as a racetrack, a young Dipper crash-lands onto the bankside. It freezes, confident that we hadn't seen it, giving us the chance to take in its dapper, scaly attire at close quarters.

Violet Oil Beetle.

JUNE

The food pass

ALONG WITH THE SKYDANCING DISPLAY FLIGHT (see *April*), the food pass is a classic and much-admired harrier trait. It takes place from early in the season as the male brings in food for the female, but it is more likely to be witnessed when nests contain young and a high number of food items are required to sustain them. This is an ingenious adaptation that has evolved because, from the nest-building stage onwards, it is the male who does most of the hunting, while the female spends more time at the nest and has a greater role in feeding the nestlings. Prey items must therefore be passed from one to the other as quickly and efficiently as possible.

In other birds of prey, such transfers would tend to take place at a handy perch close to the nest, such as a tree branch, but that is not an option for a bird nesting in open country. The transfer could be made on the ground, but bringing it directly to the nest would risk giving way the location – especially as the pale grey males are so conspicuous. Even landing away from the nest would risk leaving scent from the prey that might attract predators into the area. There would also be a chance that small prey could be lost when trying to transfer it within dense vegetation. What they do instead makes full use of the Hen Harrier's superb visual acuity, long legs and dexterity in flight: the prey is thrown by the male and deftly caught by the female in mid-air. Problem solved!

There are times when the female is reluctant to leave the nest, particularly if she is incubating eggs or brooding small chicks in bad weather. In this case there is no choice but for the male to deliver the prey directly to the nest site which he does by flying in very low and almost coming to a halt, briefly, above the nest as the food is dropped into it.

A food pass (opposite). The female flies up from the nest and deftly snatches food thrown to her by the incoming male. He can resume his hunting duties, while she feeds the nestlings. Both birds may call before the pass takes place, alerting each other that the exchange is about to be made. The male gives a low 'chuckle' and the female has a more insistent, higher-pitched 'wail'. Observations have shown that food transfers are highly efficient; prey is lost only on very rare occasions.

The 'don't think you're getting any of this', undignified scuttle across the moor.

Food pass hits and misses. Like the skydance, this was a behaviour I really wanted to study. I was not let down. When functioning it was like watching a well-rehearsed aerial display. The male would announce himself by calling, she would respond by calling and flying out to greet him. The food would be dropped as she approached from below or when flying alongside him. And with a graceful roll on her back or swoop down she would thrust out her talons (sometimes one) and grab the meal. It was a beautiful interaction to behold, though not all passes followed this script with such precision.

The 'piece of cake, I can do this', unceremonious drop straight into the nest.

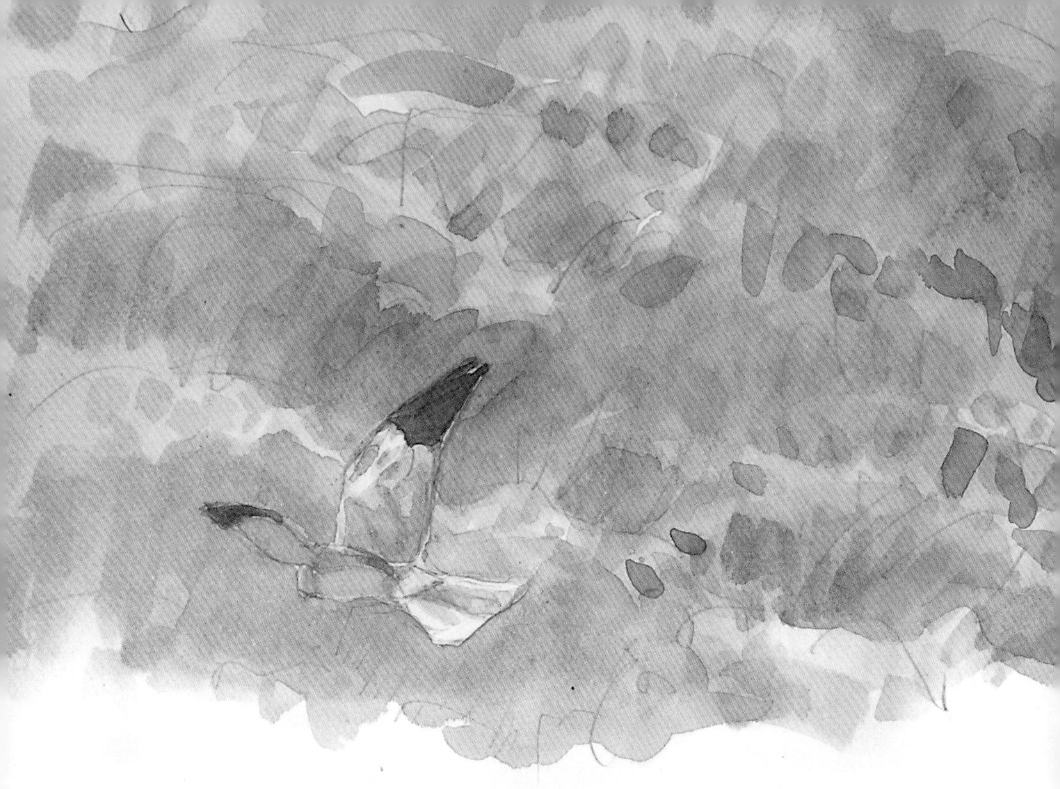

Preparing the meal. After the food pass, the female might take the food straight to the nest, the male heading back out to hunt again, or resting for a bit. At other times the male escorted her to a favoured rock, where she could tear the food into bite-size morsels before taking the meal to the waiting mouths. Or she went alone to the rock. On one occasion she sorted out breakfast mid-air – maybe having a little for herself before taking it to the young. Much chatter involved.

Setting the boundaries. A male takes time to relax, contemplating his surroundings after a successful delivery of food. The female, having none of this, flies at him, chittering away, jumps on his back, talons exposed. Not impressed by the chastisement, but not hanging around for more, the male sets off again to hunt. The same male was taking a little too much interest in the nest, frequently hovering over it and sometimes landing. A couple of angry chitters and a close fly by, sends the message that he should be elsewhere.

A different pair, a different understanding of roles. The male escorting the female back to the nest, swooping and dancing around her. Their bond further strengthened, the female heads directly to the nest and drops into it. The male goes off hunting.

A JOYOUS DANCE.
SWOOPING SKYDANCING
JUNE '10

Nestling development and brood reduction

The young nestlings grow rapidly provided the food supply is good, and they gradually become less vulnerable to predators as they increase in size and their legs and sharp talons develop into useful defensive weapons. If threatened, they back away and thrust out their long legs to present the claws. This behaviour has been known to catch out fieldworkers visiting nests to ring the young, leaving them battle-scarred on the back of their hands. It can also deter all but the most determined of predators. A nest camera in southern Scotland revealed that even an adult Fox can occasionally be persuaded to look elsewhere. The footage showed a well-grown nestling frantically flailing out with its long legs towards the snout of the intruder. The fact that it had just watched its four siblings being removed from the nest no doubt added to the intensity of its fight; the Fox, perhaps satisfied enough with the four plump chicks already taken, beat a retreat.

As with many birds of prey, if food is in short supply there is a mechanism to ensure that at least some of the young have a good chance of surviving. Because the eggs in a clutch are laid at intervals and incubation starts well before the clutch is complete, the young hatch in sequence, rather than all at once, resulting in a significant size difference between them. If food is plentiful then all the young will be well fed and continue to grow. But if the adults struggle to find food, due to prolonged spells of poor weather, or because the male is busy helping females at other nest sites, there will be a shortage. The largest nestlings push themselves forwards and grab what they need, leaving the smaller chicks to go hungry. If the situation becomes desperate, the larger nestlings will even attack the younger chicks to keep them in their place and help reduce the competition for food. Unless the food supply increases rapidly the weakest chicks will die and may become food for their older siblings. This is known as 'brood reduction' and is an excellent example of a behaviour that may seem cruel but has been honed by evolution, over countless generations, to maximise productivity. It is effective because if an inadequate food supply were to be shared equally between all the young in a large brood, it is unlikely that any of them would survive. It is worth starting off with a large brood because this offers an opportunity to fledge lots of young if food is easy to come by. But if food is limited, brood reduction ensures that at least the strongest young survive and the breeding season is not completely wasted. It's a strategy that is clever and ruthless in equal measure.

Four nestlings huddled together in the nest cup. One has almost completely lost its down and has its first true feathers – like a miniature version of the adult female. The others have new brown feathers starting to emerge in patches through the down. They are siblings, but if food is short there will be no love lost as they all battle to grab more than their fair share. The smallest of the four will only survive if there is food left over once its larger brothers and sisters have had their fill. Indeed, at one nest, the female was seen to fly off with a dead chick before returning with the headless body to present as food for the survivors.

Food in the breeding season

The diet of harriers in the breeding season has been studied mainly by looking at food remains and pellets found at nest sites. Pellets are formed from the undigested remains of prey, including fur and feathers, which are regurgitated by the adults and well-grown nestlings. It is possible to carefully break these apart and work out what prey species have been consumed. Another approach is to watch nest sites from a hide to observe what the adults bring in and, increasingly, high-tech cameras are deployed to record activities at the nest, including deliveries of prey items.

It is no coincidence that Hen Harriers have growing young in the nest at a time of year when food is most abundant in the uplands. The voles and pipits that attracted them to the breeding areas in the early summer are still present, and now have young of their own to swell the numbers. Meadow Pipits are especially important and can form up to 45 per cent of the prey fed to nestlings. They are joined by a range of other songbirds, including Skylarks, Stonechats and Wheatears that use the moors to rear their young before heading south, or to lower ground, for the winter. Harriers breeding close to young conifer plantations or patches of scrub also hunt the fringes of these habitats for woodland songbirds including Siskins, Redpolls, Willow Warblers and even Crossbills.

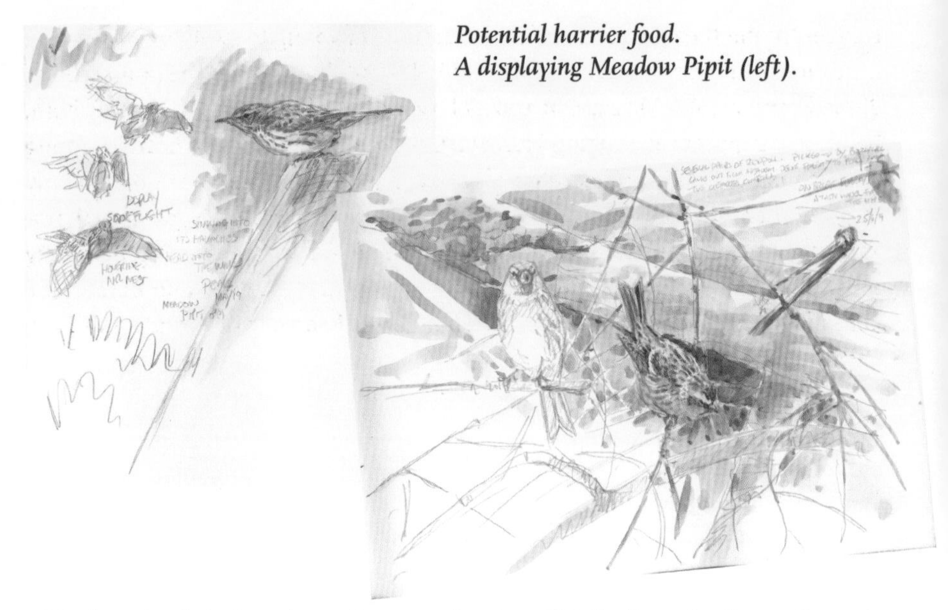

Potential harrier food.
A displaying Meadow Pipit (left).

Picked up by their buzzing calls, several pairs of Redpolls make forays from the adjacent forest edge to feed in the tangled debris of a cleared conifer block. Always wary, frequently nipping back to the safety of the dense forest.

They will also take young Woodpigeons from their flimsy nests if these are not sufficiently concealed. In Northern Ireland, pellet analysis revealed that Meadow Pipits, Starlings and Skylarks together made up 75 per cent of the total prey items at nest sites, though a wide range of other birds were also taken.

Larger birds, including Red Grouse, breeding waders and occasionally even young birds of prey, are also exploited in the breeding season. The grouse are ever-present on the moors but they are too large for the male harrier to catch easily or to transport back to the nest. Still, they produce large broods of young and these are very much on the menu. Adult grouse have been seen jumping up at the last second to avoid being caught on the ground by a stooping harrier, or to protect their young, but this defence is far from fool-proof.

Breeding waders come to the moors in the summer, notably Golden Plover, Curlew and in the wetter areas, Snipe and Lapwing. Their young, and sometimes the adults of the smaller species, attract the attention of foraging harriers. Harriers are opportunists and almost any bird that is not too large is at risk if it is surprised within foraging range of a nest site.

While birds tend to comprise the bulk of the Hen Harrier's diet in most areas, mammals are also important in the breeding season. Field Voles are taken frequently and they too produce large numbers of young to swell the population in summer. In Orkney there are no Field Voles but the Orkney Vole (a subspecies of the Common Vole) is present and its larger size makes it a valuable prey item, helping to support the thriving Hen Harrier population on these islands. Young Rabbits and hares are killed often if they are available. Rabbits are particularly important on the Isle of Man, where voles are not present. Ireland has very few voles but other small mammals including the Wood Mouse and Pygmy Shrew are taken, as are larger species including Brown Rats, Rabbits and hares. One pellet from Northern Ireland contained fur from a Badger, which could only have been the result of scavenging.

Three Snipe displaying in the air at one time – a clockwork 'tick-tock' call, diving, tail feathers whirring, drumming. A Cuckoo announces itself, calling continuously. The Redpolls carry on buzzing. All very busy... Then suddenly there is silence when a male harrier approaches overhead.

A recent study used a novel approach to look at diet in the breeding season. Swabs were taken from 52 broods of nestlings across Britain during routine nest visits by fieldworkers, and the DNA was analysed to determine the prey species consumed. In all, 62 different species of prey were detected with Meadow Pipit, Red Grouse, Wren, Skylark and voles the most frequent. As expected, the prey on grouse moors included a high incidence of grouse, but a lower than average overall species richness.

Hen Harriers share their moorland breeding areas with other birds of prey. Interactions are frequent, fuelled by a fear that small nestlings may be at risk from other predators needing to feed their own young. Sometimes, however, there are squabbles over food, and Hen Harriers often seem to have the upper hand. They have been seen stealing prey from Kestrels and Short-eared Owls – flying up to intercept them at speed and using their excellent manoeuvrability and long legs to snatch away the prize. On other occasions, interactions result in the benefits flowing the other way. Merlins have been watched circling over a hunting Hen Harrier and chasing pipits flushed, but not caught, by the larger raptor.

Sketches of Red Grouse – the bird at the centre of all the controversy. To many it's a commodity, for others a political tool. It's easy to forget that this is an engaging bird in its own right. Skulking and secretive around the nest. Great value when the males display – a heraldic emergence, glorified by a pot-bellied flap and glide, ending in an undignified crash back into the heather. It's a noisy affair too. I can't fully describe the odd noises that spill out, except that they put a lot of effort into them and the sounds raise a smile every time I hear them.

Hen Harrier diet in the breeding season in Britain and Ireland

Species group	*Notes*
Small songbirds	Meadow Pipits are often the most abundant species recorded at nests, with Skylarks and Starlings also important and a wide range of other species taken less frequently. Mostly open country birds but birds associated with woodland and scrub are taken when available locally.
Gamebirds	Red Grouse is the most important species in moorland breeding areas but at 500–600 g they are near the upper size limit for prey. Adults are killed mainly by the larger female harrier, or if sick or injured. Grouse chicks are an important food and small enough to be easily carried back to the nest. Pheasants, partridges and other grouse are taken if available, especially young birds.
Waders	Adults of the smaller species may be killed but the smaller and inexperienced young are more likely prey for harriers. Golden Plover, Curlew, Snipe and Lapwing are the species most often recorded.
Pigeons	Young Woodpigeons may be taken from their nests.
Other birds	There are occasional reports of ducks and even geese being attacked, though these birds are too large to be considered normal prey and are likely sick or injured. Small young of wildfowl are more commonly taken. Other species recorded occasionally include Corncrake, Water Rail and the young of Kestrel, Merlin and Short-eared Owl.
Small mammals	The Field Vole is a key species in mainland Britain but is absent from Orkney and other Scottish islands, the Isle of Man and Ireland. In Orkney, the larger Orkney Vole (a race of the Common Vole found on the Continent but not in mainland Britain) is an important prey species. Other small mammals including mice and shrews are sometimes taken as, occasionally, are the larger Brown Rat, Water Vole and even Hedgehog.
Lagomorphs	Rabbits, Brown Hares and Mountain Hares are all taken when available, especially the young which are more easily caught and can be carried back to nests. Rabbits are important prey on the Isle of Man where they are apparently killed and dismembered so that they can be taken back to the nestlings.
Other mammals	Hair from a Badger was found in a pellet, providing evidence that large mammals are sometimes scavenged.
Other vertebrates	Rarely taken, though Common Lizard, Slow Worm, Adder and Common Frog are all killed infrequently. Fish may be scavenged on rare occasions.
Invertebrates	Beetle wing-cases found in pellets are mainly from the Dor Beetle and ground beetles. They could be ingested incidentally with other prey but are probably also taken opportunistically when encountered. (In Spain and eastern Russia, grasshoppers, crickets and locusts form part of the diet.)

CONFLICT ON THE GROUSE MOORS

A UNIQUE SITUATION

BIRDS OF PREY ARE OFTEN PORTRAYED as a modern success story. They are a highly visible symbol of how far we have come since the dark days when organised persecution drastically reduced raptor populations across the country. As a result of persecution, we lost our Goshawks, White-tailed Eagles, Marsh Harriers and Ospreys in Britain. Red Kites, Buzzards, Peregrine Falcons and other raptors managed to survive in remote areas where persecution levels were lower, but their numbers were greatly reduced. The Hen Harrier was also heavily affected and by the end of the nineteenth century it clung desperately to survival only in Orkney and in parts of the Hebrides off the west coast of Scotland.

The success story is in the comeback that has followed stricter legal protection for our birds of prey and a decline in persecution across most of the country. The Buzzard is back as the common and widespread bird it once was, and is now our most abundant raptor. Peregrines have returned to the lowlands and now breed in our towns and cities, adopting buildings as artificial cliffs for their nest sites. Ospreys have returned and they are moving into new areas through reintroduction projects as well as by spreading naturally. White-tailed Eagles, Red Kites and Goshawks have been restored through reintroduction schemes and are expanding their ranges to recolonise long-lost haunts. We can take great pride in these achievements. They show how much mindsets have changed for the better in recent decades as birds that were once despised are welcomed back. Their presence in our countryside is rightly celebrated and our lives have surely been enriched by their return.

The Hen Harrier too has recolonised lost ground and is, once again, breeding across much of Scotland, parts of Wales and sporadically in northern England. But here the story changes. Whereas most birds of prey have either fully recovered or are on an upward trajectory, the beleaguered Hen Harrier remains scarce and has been unable to make significant gains. In England, only a few pairs breed in most years, and across vast expanses of moorland that should support them, they remain absent or very rare. This is the direct result of conflict between the Hen Harrier and a small upland gamebird, highly esteemed for the sport it provides to shooting enthusiasts. The reception awaiting Hen Harriers that visit our grouse moors provides a modern-day insight into Victorian attitudes. You may have thought they had been consigned to history but, if you look in the right places, they are alive and well. This has become a hotly debated and increasingly divisive issue, with conservationists desperate to ensure that the Hen Harrier is protected, and devotees of driven grouse shooting focused on defending their sport.

Driven grouse shooting

This is a form of hunting that is unique to Britain. It takes place on moors specifically managed to encourage high densities of Red Grouse: a race of grouse found nowhere else in the world. The grouse are not artificially reared and released as happens with Red-legged Partridges and Pheasants in the lowlands. But the moors are carefully managed so that conditions are perfect for them. Patches of heather are burnt in rotation to ensure that there is always plenty of fresh young growth for them to feed on. Older, unburnt patches provide cover for nests and shelter so that this hardy bird can see out the toughest of winter weather. Predators are controlled ruthlessly and the grouse are even provided with medication, in plastic trays dotted across the moor, to help control the

A walk across a grouse moor in late summer will reveal the unnaturally high densities of Red Grouse but perhaps not much else for the wildlife enthusiast. During the season, from the 'glorious' 12 August to 10 December, shooting takes place on moors across the north of England and parts of Scotland. Beaters push coveys of birds towards a line of guns, crouched in the grouse butts, expectantly waiting for the grouse to come hurtling towards them. A Hen Harrier is not a welcome sight for the moor owner or shooters. All the time it is present on the moor it is likely to be eating into the grouse stocks. And its low hunting flights can spook the grouse, or scare them away, making it more difficult to drive them over the guns.

Plastic posts litter the moor, marking the location of trays containing medicated grit so the gamekeepers can find them again.

spread of disease. As a result, grouse are able to increase to the astonishingly high artificial densities required for driven shooting.

During each drive, groups of Red Grouse are flushed by a line of noisy, flag-waving beaters and pushed towards the guns. The sport is fast and furious and the guns, concealed in their grouse butts, expect large numbers of birds to be presented as targets (the record is 2,929 grouse shot on a single day – by a party of just eight people). It is an expensive day out. On the most highly esteemed moors a day's shooting might cost £2,000 or more for each person. Participants demand good value for money and plenty of grouse to shoot at. Enter the graceful, elegant and unwelcome Hen Harrier, sweeping over the moors on the hunt for pipits, voles and perhaps even a Red Grouse or two.

The role of science and technology

For a long time conservationists argued that although Red Grouse were sometimes killed by harriers, this would have little or no impact on the overall grouse population on the moors. But they were wrong. While it's true that many predators have little impact on the overall population of their prey, grouse moors are rather different. Studies showed that in order to maintain the extremely high densities of grouse required to make driven shooting viable, almost all predators must be eliminated. Gamekeepers are rigorous in ridding the moors of Foxes, Stoats and corvids, all of which can be done legally by trapping, snaring and shooting. Hen Harriers and other raptors are legally protected. But if they are left to thrive, the research shows that they can make a real dent in grouse numbers and, in some cases, driven shooting would no longer be viable. The harsh reality is that this sport can only be maintained if gamekeepers act against all predators, including the Hen Harrier.

Of course, not everyone involved in managing our uplands resorts to illegal persecution. Here too, the science is instructive. Studies using radio and satellite tags fitted to harriers have shown that they roam across many different areas before they settle to breed. If persecution is carried out across a high enough proportion of the uplands then sufficient birds will be lost to prevent the population from increasing. And that is exactly what happens. Tracked birds wander widely, but a recent study based on a large sample of satellite-tagged Hen Harriers in England revealed that birds were a staggering ten times more likely to die or go missing (presumed dead) in areas managed for grouse shooting. Modern tags are carried by an increasing number of harriers and the association between grouse moors and illegal persecution has become all too clear in recent years.

Upland estates where illegal activities are not undertaken, including some managed as nature reserves, struggle to support harriers because so many birds are killed when they move away from these safe havens to explore the surrounding landscapes. A nest site on a nature reserve will fail if the adults are killed when they stray beyond its boundaries. Often the adults simply vanish part way through the breeding season – they fly out from the nest on a hunting trip, never to return. Their fate is rarely witnessed because of the remote nature of these landscapes and a paucity of witnesses on private estates.

Scientists have been able to estimate the overall impacts of illegal persecution. An early study involved visiting nests and fitting coloured wing-tags to over 1,600 young birds, mainly in Scotland, over a number of years. Based on re-sightings of these birds once they reached adulthood, it was shown that survival rates were far lower on grouse moors than for other habitats. Overall, an estimated 11–15 per cent of the breeding females in mainland Scotland were killed by gamekeepers every year, an annual illegal cull of 55–74 adult breeding birds. In addition, many of the nests on grouse moors were destroyed.

By looking at the various factors that influence the distribution of harriers in Britain and then assessing the amount of habitat available, scientists have worked out how many birds our uplands could support. The conclusions are startling. In Scotland there is enough habitat for over 1,500 pairs of Hen Harrier and yet the last survey found just 460 pairs. In England, upland habitat is less extensive but the discrepancy is even more stark. In 2021 there were 24 successful pairs in northern England, but there is enough habitat (in the uplands alone) to support an estimated 323–340 pairs. In Northern Ireland and Wales too, the current populations represent a small fraction of what the habitat should be able to support. The following section looks at this problem in more detail and describes the different forms of persecution enacted in order to keep the Hen Harrier out of large parts of our uplands.

The killing fields

It is often thought that wildlife crime is primarily a problem for other countries to deal with. You might think of the illegal poaching of large mammals in Africa or the trapping of migrant birds in spring as they pass through countries surrounding the Mediterranean. But we must also look closer to home. Normally, when Hen Harriers are killed, it happens in remote upland areas and those involved will conceal the evidence, aware of the stiff penalties should they be caught. If tagged birds are killed, the data will show where they were lost, but if the body is hidden and the tag destroyed then the precise details of what happened to the bird will never be known. Yet, against all the odds, a few incidents do come to light and they reveal the variety of methods employed by those who wish this bird harm.

The Hen Harrier is particularly vulnerable during its long breeding season when its eggs and young are hidden at a nest on the ground and so are all too easily accessible to humans. If the nest is discovered by someone fearful of the impacts on grouse, it is likely to be destroyed. Fieldworkers have even found well-grown young dead in the nest, flattened by the trampling of a heavy boot; all the effort of the season so far comes to nothing as another breeding attempt ends in failure. Adults are also vulnerable once their nest has been found and gamekeepers have been watched lying hidden in the heather waiting for them to return.

Incidents of illegal shooting have been recorded by miniature cameras installed by conservation workers as part of their efforts to monitor activity at the nest. One recent incident involved an armed individual hiding near to an artificial decoy bird, painted to look like a male Hen Harrier and placed out in the open

Tiny, covert cameras are increasingly used to record what happens at Hen Harrier nest sites. Even when they have recorded an adult bird being shot as it flies up from the nest, it has proved impossible to gather enough evidence to successfully prosecute the individual involved. But such individuals tend to be dressed as gamekeepers and the crimes take place on private grouse moors managed for shooting. These crimes are there for all to see on video-sharing websites, testament to the ongoing threats that Hen Harriers face as they try to rear their young.

on a grouse moor. The aim was presumably to attract in a male bird deceived into thinking the decoy was a rival, providing the opportunity to shoot it.

A few years ago, RSPB staff went to great lengths to try to protect a nest near the edge of their nature reserve at Geltsdale in Cumbria, where it borders a neighbouring grouse moor. Staff would arrive before first light and hide in the heather to watch over the nest area. One morning, at 05.37, a loud shot was heard, soon followed by a second blast. An individual with a full-face balaclava appeared to retrieve something from the heather and hide it. Once he had left, the site was searched and a dead bird pulled from a hole at the edge of a drainage ditch. All the protection efforts had been in vain; it was the body of a breeding female. The police were involved and took the investigation seriously but, as is so often the case, it proved impossible to identify the individual responsible.

Shooting is also a problem in winter when birds gather together at traditional communal roost sites in the late afternoon. These can be staked out and the birds shot as they fly in to seek a secure place on the ground to spend the night. Opportunistic shooting takes place throughout the year as and when a bird is encountered at close range. Here too, modern technology can help reveal the crimes that are being committed even if it does not help to catch the perpetrator. Occasionally, the bodies of birds that have been shot and wounded are later recovered by following up the signal emitted by their high-tech tags. But technology of a different kind is also helping to facilitate the killing. Firearms with telescopic sights and advanced night vision equipment means that birds can be killed from a considerable distance, even under cover of darkness.

Hen Harriers, along with other birds of prey, sometimes use perches such as fence posts to rest and survey their surroundings before deciding on their next move. This fact can be exploited by placing a small trap on the top of the post. The jaws of the trap snap shut using a powerful spring mechanism when anything lands on the central trigger plate. These traps are used legally under cover to catch Stoats and rats. When set out in the open they are illegal and those who have seen the results will understand why. A bird landing on the post has its legs crushed and is then held in place in the jaws of the trap until it dies a slow death, flapping desperately to escape until exhausted. Another raptor successfully removed from the moor and no longer able to threaten the local grouse.

The use of illegal poison baits is most effective against species that regularly scavenge on dead animals, including the Red Kite, Buzzard and, in Scotland, both species of eagle. Harriers usually hunt for their own prey but they may be tempted by small baits if these are placed close enough to their nest. This has occasionally been used as a tactic against them and, once again, Geltsdale,

The story of Bowland Beth

Bowland Beth, a Hen Harrier named after the place where she was reared, flew to a grouse moor in Nidderdale, North Yorkshire where she was shot and killed. She managed to evade her killer by flying on after being wounded but died soon afterwards. Her body was found by tracking the signal from her satellite tag and a post-mortem was carried out to confirm the cause of death. This is a fate that befalls many of our Hen Harriers but because, in this case, the body of an individually known bird was found, it received considerable publicity and helped to increase public awareness of the Hen Harrier's plight. It even resulted in a book, *Bowland Beth: The Life of an English Hen Harrier* by the late David Cobham, a respected wildlife film-maker. In it, he pieces together the story of her short life using data from her satellite tag. Towards the end of David's book is this quote from Stephen Murphy, Natural England's Hen Harrier Recovery Officer:

> *Beth was a beautiful bird – an amazing bird – and I feel so privileged to be the only human to have held her while she was still just a bundle of earthbound feathers and attitude. Her story is remarkable. We should be celebrating her life now and her becoming a parent, and tracking her sons and daughters.*

Bowland Beth at Langholm in southern Scotland.

Overly sentimental for just one individual bird you might think. And yet Stephen had monitored the nest where she was reared, visited it to fit her with a tag when she was a few weeks old, and wished her well as one of only a handful of young Hen Harriers to have flown from nests in England that summer. It was Stephen who followed her progress day by day over the next year as the tag revealed the pattern of her wanderings back and forth across the uplands. She ventured to the north of Scotland and then returned in spring to the Bowland Fells, close to where she was born, in an unsuccessful attempt to find a mate. Finally, after another excursion to the far north of Scotland, she drifted back south, ending up on a grouse moor in North Yorkshire. And here, it was Stephen who tracked her down to her final resting place in a patch of heather. He held the bird in his hands for a second time, now nothing more than a limp, lifeless body with a shattered leg and Stephen's tag still attached. Just one bird but a vivid reminder of why harriers are missing from so much of our uplands. Perhaps not overly sentimental after all.

an RSPB nature reserve hemmed in by grouse moors, was the setting for an incident that was discovered by fieldworkers. There were two nests on the reserve that year but both failed when the adults went missing. A search located a dead female and analysis confirmed that she had been poisoned with the banned pesticide mevinphos. A subsequent search found several Starlings laid out in the open along a fence just inside the reserve and close to one of the harrier nests. Tests confirmed that they had been laced with the same poison.

In another incident, no fewer than 32 pieces of poisoned meat were found along the boundary fence of a Scottish grouse moor, and in North Yorkshire, thirteen 1-day-old poultry chicks laced with mevinphos were recovered from posts. There can be no doubt the majority of such incidents, taking place in remote upland areas, go unrecorded.

A final technique worth mentioning involves less in the way of blood-letting but can be highly effective in preventing harriers from settling, or moving them on if they are trying to establish a nest site. This involves removing suitable patches of Bracken or tall heather that would otherwise provide good cover for a nest. In areas where heather burning is intensive, these patches of taller vegetation are in short supply so the harriers have few other options if their chosen site is destroyed by cutting or burning. If harriers are seen trying to settle at a likely breeding site, deliberate disturbance can also be used to persuade them to move on. These subtler techniques seem to be on the rise as those involved with persecution grow increasingly fearful that direct illegal killing or nest destruction will be witnessed or caught by covert cameras.

Possible solutions and a way forward?

The field of conflict resolution is about bringing together groups with opposing views in order to agree a compromise that will satisfy everyone. Much time has been invested in this approach to try to reduce the conflict between Hen Harriers and grouse shooting, partly funded by a government keen to restore harmony to our uplands. And in a clear demonstration of the intractability of this issue, all these efforts have come to nothing. If anything, views have become even more polarised and the two camps are now further apart than ever. Before coming on to consider the future, it is worth reflecting on some of the solutions that have so far been proposed. Firstly, two proposals aimed at reducing the impacts of harriers on grouse and so reducing the incentive to kill them. Then a third option to try to improve the Hen Harrier's fortunes by establishing a new population well away from the moors.

Diversionary feeding

This involves the provision of food on specially built, raised platforms close to Hen Harrier nests. A constant supply of poultry chicks or rats ensures that the nestlings are well fed and, as a result, the adults are less inclined to roam the moors in search of food, including grouse. The approach has been trialled as part of an in-depth study at Langholm Moor in southern Scotland. And it works, at least up to a point. Adult harriers did indeed bring fewer grouse to the nest when they had an all-day buffet available close to their nest. The problem with this technique in the minds of moor owners is one of practicality and expense. If harrier populations are left to build up then each grouse moor would host an ever-increasing number of active nests. Installing feeding platforms and keeping them well stocked with food during the rearing season would become a significant undertaking and would need to be carried out every summer. There

Hen Harriers hunt for their prey and do not normally scavenge in the manner of a Red Kite or Buzzard. But if dead animals are provided on a prominent platform close to their nest, they will use them to feed the nestlings. This reduces the need for hunting and means that fewer grouse will be taken from the surrounding area.

are also concerns about attracting other scavengers such as corvids onto the moor, and that the increased number of young Hen Harriers fledging from nests could cause disruption on shoot days by spooking the grouse. Despite the initial promise, it seems there is not much appetite among moor owners to pursue this option.

Quotas and brood management

This option is rather more popular with moor owners. The idea is that on each grouse moor a small number of pairs are allowed to settle and breed unmolested, once a 'quota' (based on the size of the moor) has been agreed. When the quota of nests is reached, further pairs are prevented from settling – or if they do settle, the eggs or young are taken into captivity under licence for rearing and release elsewhere. Moor owners are thus reassured that the harrier population will not increase to a level that could prevent driven grouse shooting and so, it is hoped, become more relaxed about hosting a small number of birds.

This approach is expensive and it involves the unprecedented step of artificially limiting the population of one of our rarest birds of prey. In fact, some conservationists were so horrified by the idea that they challenged its legality in the courts. The legal position has yet to be fully resolved but, in the meantime, a brood management trial started in northern England in 2018. Each year, when the threshold of nests is reached on a moor, chicks are taken into captivity and released elsewhere in the uplands later in the summer. This measure was supposed to reassure grouse moor managers that the Hen Harrier population could not increase unconstrained. As a result, illegal persecution would no longer be necessary. So, is it working? Well, there are encouraging signs of a gradual upturn in the harrier population from a low point of just 4 pairs in 2016 to 24 successful pairs in 2021. And yet illegal persecution continues to take place apparently unrestrained. Since 2018, more than 50 birds have been killed or have disappeared in suspicious circumstances (based on reliable data from their tags). The illegal activity that brood management was supposed to curtail is still occurring.

A lowland reintroduction

While this is not a solution to the problems in the uplands, it would help to improve the overall status of the Hen Harrier in Britain by reintroducing it to the lowland habitats in southern England where it once bred. This approach has worked very well for the Red Kite, Osprey and White-tailed Eagle and it would help to establish Hen Harriers in an area well away from the persecution hotspots on the grouse moors. It is seen as a positive way forward by grouse

Harriers once nested in the lowlands. Could they make a return to our lowland heaths, downs and farmland, away from persecution on the moors – or is that a dream too far?

moor owners, perhaps because they feel it would take some of the attention away from problems in the uplands. Some conservationists are supportive but others are concerned that a reintroduction involves too much unnecessary human intervention when all that is really needed is for illegal activities to cease. The debate continues and it is unclear whether this project will be taken forward.

The way forward?

There are other possible measures that could help the beleaguered Hen Harrier. One hope is that our enforcement agencies will put more effort into trying to catch and punish the wildlife criminals that are so rarely brought to justice. In remote areas that is far from an easy task. But ever-improving technology, including more accurate and longer-lasting tags, as well as increasingly sophisticated covert cameras, will certainly make it easier to gather evidence that stands up in court. Another option that would work alongside enforcement is the tighter regulation of driven grouse shooting through a licensing scheme. The licence required to operate could be revoked if illegal activities were found to have taken place. This option has been agreed by the Scottish Government, although, at the time of writing, it has yet to be put in place.

One change already implemented in Scotland is so-called vicarious liability, in which the owners and managers of land are liable to prosecution if wildlife crimes take place on their estates, even if they are not the ones carrying out the activity. The hope is that owners will be more careful about the expectations of their employees if they are the ones that could, ultimately, be prosecuted. It would certainly help if this was introduced in England though it would only be effective if we become better at detecting crimes and the individuals responsible for committing them.

Despite all the proposed solutions (as yet without much in the way of meaningful progress) there is a growing feeling that driven grouse shooting is simply incompatible with sensible management of the uplands. This view has gained support in recent years amid mounting frustration over the continued illegal killing of Hen Harriers and other species. Jeff Watson, in his book *The Golden Eagle*, was as concerned about the impacts of illegal persecution on his chosen species as his father Donald had been about the impacts on harriers. He includes a quote from his father's book, *The Hen Harrier*, in which Donald reflects on the attitudes of some grouse moor owners, concluding that, to them, 'the law is somehow an impertinence on the moorlands of Britain'. That was true when *The Hen Harrier* was first published in 1977. It was still relevant when *The Golden Eagle* was published in 2010. And it remains true today. Sadly, neither Donald nor his son Jeff lived to see significant progress made.

In addition to concerns about persecution, there are wider issues connected to the intensive management of grouse moors. Grouse shooting has spared some sites from the twin horrors of overgrazing by sheep or planting with dense, serried ranks of non-native conifers. The intensive burning carried out on grouse moors maintains habitat that is used by the grouse themselves and also by ground-nesting waders such as Golden Plover and Curlew. Yet most other wildlife loses out and moors are often rather soulless and dispiriting places for those who take the trouble to visit them. In autumn, once the waders and other summer visitors are gone, it's possible to walk for hours and see few birds other than the ubiquitous grouse and perhaps a few Meadow Pipits. The regular burning and other management causes pollution, releases carbon into the atmosphere, damages sensitive wetland habitats and, as we have seen, removes predators wholesale using legal and often illegal methods.

Another concern, much in the news recently, is the effect that burning the vegetation has on the flood risk for communities in the valleys below the moors. Bare ground and short vegetation allows water to run off more quickly. Studies have suggested that the likelihood of flooding is increased and recent damaging floods at Hebden Bridge in North Yorkshire and elsewhere have been attributed, at least in part, to management on the moors above.

Given these wider concerns and the ongoing problems highlighted by the Hen Harrier, perhaps the time has come to consider alternative options for managing parts of our uplands. Could some of them be 'rewilded', allowing a greater range of vegetation to thrive, including bushes and trees? This would attract a far wider variety of wildlife, as well as visitors wishing to come and see it. Carbon would be locked into the vegetation, flood risk would be reduced and unhealthy plumes of smoke would no longer drift across the moors into nearby communities. Something far less artificial and constrained, and far less damaging to the wider community could be restored for all to enjoy.

Drifting too close to a Merlin's nest, the male harrier flushes out a distraught male falcon, who makes a speedy stoop at him – sensibly he left it at that, but continued to have a lot to say. The harrier hardly seemed to notice the kerfuffle it had caused and carried on working the clough in its deceptively lazy manner, casting ever-changing shadows upon the hillside.

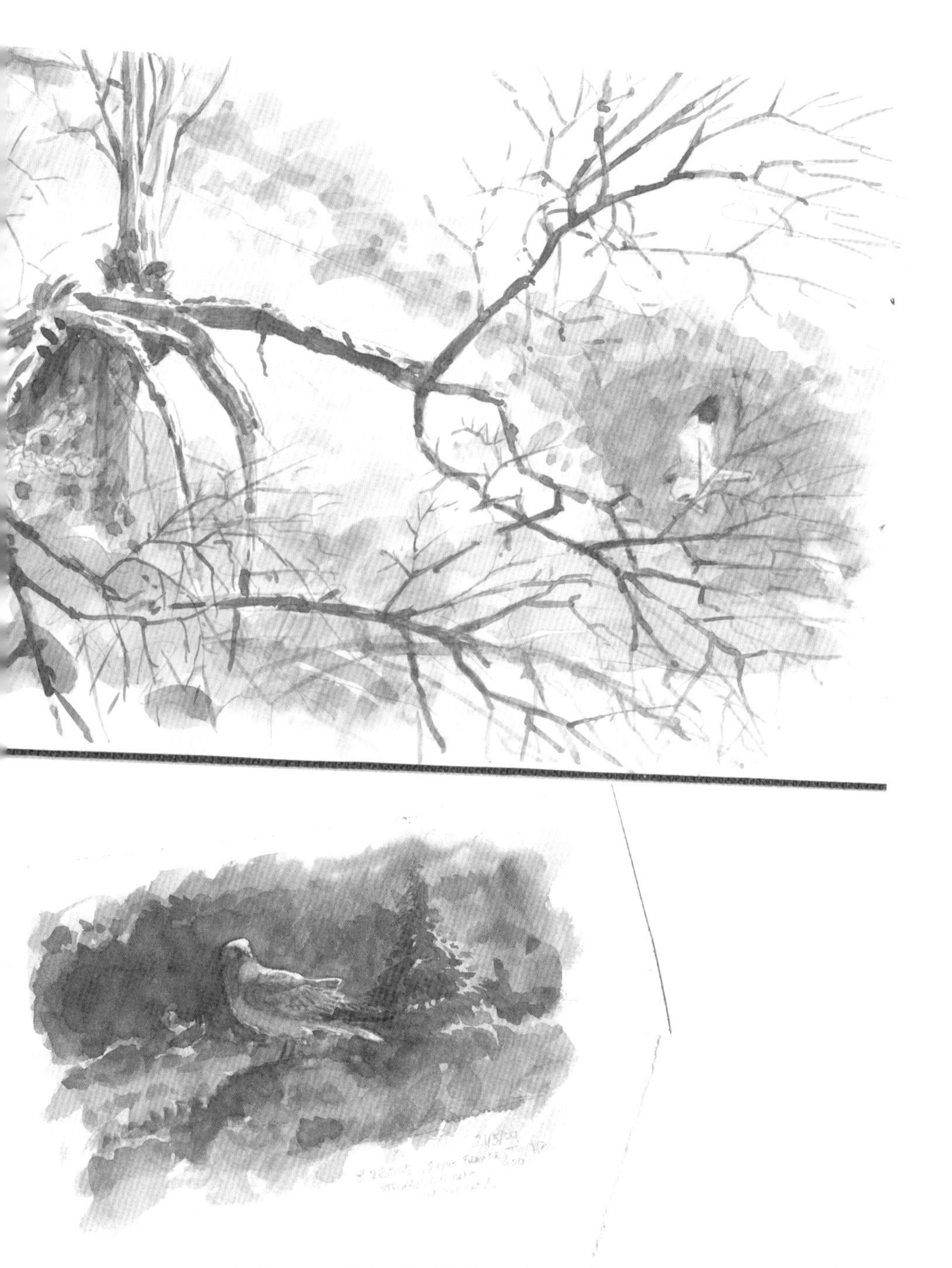

A chance to sketch the birds as they took time away from nest and hunting duties. The female often left the nest to rest on favourite rocks – deceptively relaxed, but always alert to threats.

It may be serendipity, but when moments happen all you can do is sit back and let them unfold in front of you. A pair of Ravens appeared over our shoulders, one rolling over in display to the other – 'cronk'. The harriers rise up to keep an eye on things – another shape in the distance reveals itself to be a Red Kite. Ok this is good. Then a Peregrine drifts in, casually circling the valley. Wow, what next? Well, an Osprey flaps into view! Reaches the head of the valley and wanders back down again. Not much else to say really.

A Ring Ouzel had been singing all morning, but was proving difficult to locate. Eventually seen moving around in last year's dead Bracken, using low clumps as song posts – not on the rocky crags where I had been searching for it. It was perilously close to the nesting harriers and Merlins. A different Merlin takes a fast, direct route over the valley carrying prey – avoiding a dispute with the locals.

ALL MORNING - PROVED DIFFICULT TO LOCATE,
SEEN MOVING AROUND LAST YEAR'S BRACKEN
LOW CLUMPS OF GRASS AS SONG POSTS -
CLOSE TO THE MERLIN
WHITE PATCH ON
EAR COVERT.
♂ RING OUZEL
'torquatus'
21/3/18

JULY

THE BREEDING SEASON IS NOW WELL ADVANCED and food availability is peaking just at the right time to fuel the growth of young harriers developing in their nests. Young birds and mammals are making their way in the world and their inexperience and naivety make them relatively easy prey. The days are long and in good weather there is ample time to hunt for food. The adults must make the best of these conditions while they last. Already, by the end of the month, prey availability starts to fall away and late nesting pairs face a race against time if their young are to do well and fledge successfully.

The growing brood

As the nestlings develop, they grow their first set of true feathers and become increasingly like miniature, untidy versions of the adult female. They are now less vulnerable to predators, and are better able to withstand poor conditions should the midsummer weather take a turn for the worse. They are also much more mobile on the nest, shuffling around constantly, helping themselves to food items when they feel hungry, and eagerly grabbing at any new food brought in by the female.

A few weeks after hatching, the nest cup has become a rather cramped space and the nestlings start to investigate their immediate surroundings, even creating small tunnels into the adjacent vegetation. In part this could be a defensive strategy to minimise the risk from predators should the nest be discovered. The male chicks tend to spend more time away from the nest, perhaps because they are smaller and so have less chance of defending themselves effectively. If a predator does find the nest, at least some of the young may be away from the nest cup and so will have a chance of avoiding detection. This behaviour also helps to prevent the nest from becoming too heavily fouled, though from an

Left: The male harrier's pale underparts may help it blend into the sky as viewed by potential prey on the ground. Black wing-tips break up the bird's outline, and the long 'fingers' reduce turbulence (and noise that might alert prey) and improve flight efficiency.

Well-grown nestlings are not the prettiest birds, with their broad, flattened heads, seemingly over-long legs, large feet and bits of dirty-white down sticking out haphazardly from the brown feathers. They are also rather ungainly creatures as they shuffle around on the nest platform but they have a certain character and charm all of their own.

early age the chicks make an effort to deposit their droppings beyond the edge of the nest. Raptor workers visiting the site to ring the young must be diligent to ensure that all nestlings are accounted for, rather than just examining the small nest cup itself.

Hen Harrier nestlings can be rather feisty when handled. The wings are raised in threat, the beak opens (tongue lolling out), while the long legs, tipped with sharp claws, are thrust forwards as their main defensive weapon. By about three weeks the female chicks stand out as noticeably larger and bulkier than the males, with thicker-set and longer legs. The eyes of males tend to have a grey tinge, being browner in the females. But it is the size of the feet that is usually the clinching feature. If the span of the claws is 62 millimetres or more, they belong to a female. If it is less than 60 millimetres, then the feet are those of a male bird.

Foraging range from the nest

Once the young are about two weeks old, the female can leave them unattended, at least in good weather, and start to help the male hunt for food. This is just as well because the number of prey deliveries required increases when there are well-grown young in the nest. A rough average of three small prey items are needed every day for each nestling. Adults with a brood of four or five young will therefore need to bring back a prey item every hour or so during daylight to ensure that they are all well fed.

With the female actively hunting, the range of species that can be taken increases because she can kill and carry heavier prey that will generate more food for the growing brood. If her partner has several females to provide for, it may now fall to her to find the bulk of the food for her own young. This can be difficult when food is in short supply or if there is a spell of cool, wet weather. This not only makes hunting more difficult but it means the female is required at the nest to brood the young; young up to about three weeks old still benefit from shelter provided by the female to help keep them warm and dry during spells of poor weather. The adult birds defend a small area around their nest from other harriers, but the far larger home range, where they hunt, is not defended and often overlaps with harriers from other nests nearby.

Females tend to range less widely from the nest than males, even when the young are old enough to be left unattended. A study of moorland nests in Scotland found that the average home range for males was 7.3 square kilometres but just 3.6 square kilometres for females. Males often range several kilometres away from the nest, and on rare occasions marked birds have been seen as far as 10 kilometres away as they pursue their relentless quest for food. Much depends on the quality of the food supply near to the nest, with longer trips undertaken only if they are necessary to improve hunting success. Females also range several kilometres from the nest at times but they do so less frequently and most of their hunting is confined within a radius of 1 kilometre or less. Even with well-grown young, it is as well for the larger female to remain close enough to the nest so that she can return quickly should a potential predator stray too close.

The striking colours of a male should make following his progress against the moor a straightforward exercise and generally it is ... even so, with a twist of his tail he can disappear in an instant.

Sketching a male while he rests close to the nest, I take my eye off him to make some marks in the sketchbook and in a blink he is gone. I pick him up again, already some distance away from the nest, flying purposely along the top of a clough. In a change of gear he plunges at pace, careering down low against the clough. Nothing is flushed out. When he reaches the bottom, the brakes are engaged, slowing him back down to cruise mode, before he makes his way along the brook to hunt further afield. Later he returns with prey at the top of the moor, transferring it with a perfect food pass.

It is interesting to speculate on how foraging birds make their decisions when setting off on a hunting trip. Should they keep on hunting within a kilometre or two of the nest, even if they have had little success, so that the trip back, carrying food, is not too onerous? Or is it time to start exploring further afield in the hope of finding areas richer in prey? A mathematician armed with information on hunting success, and weights and availability of prey in different areas, might struggle to work out the most efficient behaviour, and yet the harriers somehow do it all by instinct.

Moult

The regular replacement of feathers is essential in all birds, but the timing is tricky, especially in larger birds for which the process takes several months to complete. It is doubly tricky for a bird of prey that relies on its superb flying skills in order to hunt for food. Those skills will not be quite so good when some of the feathers are being regrown. Moult needs to take place, as far as is possible, at a time when food is easy to come by and conditions are good for hunting. The female harriers start their moult when incubating eggs, taking advantage of a stretch of several weeks when they are not required to fly very much. The males are very busy during this period, bringing in food for the female and then the young nestlings. For that reason, they delay their moult until later in the season, when the nestlings are a few weeks old and the female is able to help provision them.

In both sexes the moult has usually been completed by the end of the autumn so that when the more challenging winter conditions arrive (and for some the need to move to a new area), they are fully winged. First-year birds and non-breeders start the moult earlier and it progresses more rapidly because they do not face the same challenges of breeding birds with young to rear.

Moult progresses slowly so that only a few tail or wing feathers are regrowing at any one time and flight is not unduly hampered. It is not feasible to adopt the approach of some waterbirds and songbirds that replace all their feathers quickly in late summer, compromising flying ability for a short time. Nevertheless, a slight gap at the same place in each wing may be noticeable as a moulting harrier passes directly overhead.

MAGNIFICENT, DEFIANT
THE SHORT-EARED OWL
WATCHES OVER HIS DOMAIN

AUGUST

THE END OF THE BREEDING SEASON is now in sight and, already, the nights are starting to draw in and there is less time available for hunting. In a good year, there are still plenty of young, inexperienced animals available as prey just as the young harriers begin to explore their wider world. For the overworked adults, the season is almost done with for another year and the young of the early breeding pairs are already flying strongly. But for pairs that started a little later, one last push is required to ensure that their young fledge successfully and have the best possible start in life once they leave the nest.

First flight

The young nestlings spend more time standing up and flapping their wings on the nest as the time approaches for them to leave. Their muscles develop and they grow stronger with each day. The nest itself can be a rather messy and unsavoury place (at least from a human perspective) at this stage, with prey remains strewn around, attracting flies and other carrion-feeding invertebrates. Donald Watson visited a nest in southern Scotland where the eyes of the nestlings were almost completely obscured by a mass of tiny flies. Little wonder, then, that the nestlings are keen to leave as soon as they are able. It is the males that tend to leave first, usually about 30 days after hatching. The larger females take a few days longer. But if the nestlings are underfed due to food shortages, they may require an extra week or more before they are ready for their first flight.

Once the young harriers can fly, they have an escape mechanism for avoiding ground predators and their overall chances of survival are greatly increased. They could still be taken by surprise when roosting overnight, or if ambushed on the ground during the day, but this is far less likely than when they were confined to the nest. They are also less vulnerable to opportunist aerial predators such as corvids, but the threat from large or powerful birds of prey remains. In fact, despite their drab, brown plumage, they probably become more obvious as targets as they make their first, hesitant flights around the nest area. In Britain, it is probably only Peregrines, Golden Eagles and, in areas close to forests, Goshawks that present a real threat. They often patrol at height and a surprise

MAY 2019

In the south where Peregrines are a welcome addition to urban wildlife, churches and tall office blocks make for perfect 'cliff-top' nest sites. It was nice to sketch these powerful falcons in a more traditional setting. As the female Peregrine stands guard over her tiny chicks, the male patrols their territory.

attack from above could result in the loss of a young harrier before it is even aware of the threat. Such attacks are rarely witnessed but wing-tags from young harriers have been found at Peregrine nest sites in England, showing that the threat is very real.

Towards independence

In the initial weeks after their first flight, the young remain more or less dependent on the adults, primarily the female as the male is often absent by this stage. They are learning to fly properly, honing their skills and discovering how best to find food for themselves. Much of the basic behaviours they rely on are instinctive but they surely benefit from experience and practice as they refine their techniques. The youngsters frequently interact with each other in flight in what might seem like pointless frivolity but no doubt serves to further develop their aerial skills. It is their equivalent of the rough-and-tumble play that is so well known in young mammals. During this time the female keeps bringing in prey, delivering it to the old nest platform or dropping it to the young for them to try to catch before it hits the ground. The young birds have a distinctive squealing call when they are trying to solicit food, or if they become alarmed by a potential predator.

A recently fledged harrier circles close to its nest site.

Recently fledged young are darker than the otherwise similar adult females, with richer colours and a slightly different shape due to their shorter, more rounded wings and shorter tail. The underparts have an orange-brown wash beneath the darker streaks and they have a more contrasting head pattern with a dark crown and a dark patch on the cheek. Rufous tips to the wing-coverts show as a distinctive line across the upperwing. These differences become more subtle later in the year and most birdwatchers resort to the term 'ringtail' to refer to brown birds that could be either juveniles or adult females.

Breeding productivity

Overall, at least 20–30 per cent of breeding attempts in an average year will end in failure, because of either predation or a shortage of suitable prey. Failure rates can be even higher than this. A study on Skye in the Hebrides, carried out over more than a decade, found that 53 per cent of nests failed, with Fox predation identified as the most significant cause. Brood reduction provides a mechanism to ensure that at least some young survive if food is limited or poor weather reduces the time available for hunting. But in a very poor year, the adults may have no choice but to abandon their breeding attempt altogether in order to find enough food to keep themselves alive. That is a better strategy than going short of food, thus compromising their own welfare, and the chance to make another breeding attempt in the following year when conditions may be more favourable.

Polygynous nests involving subordinate females have a higher risk of failure because there is little or no support from the male bird. Once a nest has failed, the adults usually disappear from the area. They will only make another breeding attempt if the nest is lost at an early stage in the season, while there is still time to start all over again. But if they lose eggs that have been incubated for some time, or chicks, there is no option but to wait for another year to try again.

To balance the poor years, there are years when the hills are alive with voles and pipits, food is abundant and the sun shines for long hours each day through the breeding season, allowing uninterrupted hunting. Predation is less likely to be a problem because generalist predators such as Foxes, Stoats and corvids also have a plentiful food supply and so will be less inclined to risk the wrath of an angry female harrier. And if they do try their luck, the female is more likely to be close by and able to defend the young rather than kilometres away, desperately searching for prey. In these good years the most successful pairs can rear five or six healthy young and more of the low-ranking females with polygynous males will be successful.

Birds of a feather – Hen Harrier and Short-eared Owl

The Hen Harrier and the slightly smaller Short-eared Owl are often found in the same habitats and share a remarkable number of characteristics, despite not being closely related. They provide an excellent example of convergent evolution, where the same beneficial features have evolved in species that share a similar lifestyle and occupy a similar environment.

The Hen Harrier and the Short-eared Owl are perhaps roughly equally appreciated by birdwatchers. You may have your own personal favourite but they both, undoubtedly, add something special to a day's birding. And they have a surprising number of similarities. Their mode of hunting, low over the ground, is very similar. They are both active during daylight hours (which is unusual within the owl family), and they both take many voles when these are abundant, but utilise a wide range of other prey if necessary. In Britain, they breed primarily on moorland in the uplands, with nests in the lowlands far less frequent, and they move to lower ground, often favouring coastal sites, in winter. Their respective maps in the most recent national bird atlas are strikingly alike in the breeding season as well as in winter.

Short-eared Owls and Hen Harriers tend to occur at low densities when hunting, so you can expect to encounter only an occasional bird during casual birdwatching. But they roost communally in winter, so may be seen in small groups at the beginning and end of each day. Observations suggest that they sometimes even share the same communal roost sites, with the Short-eared Owls using them in the daytime once the harriers have left. Short-eared Owl pellets have been found, together with those of the Hen Harrier, on

the flattened patches of vegetation within a roost in southern Scotland. The same association has been witnessed between the Short-eared Owl and the Northern Harrier in North America.

In the summer both harrier and owl tend to breed at low density and can be highly territorial if intruders of the same species approach the nest area. But in an area where prey is abundant, clusters of nests can be found as birds home in on a place with a rich food supply. The respective breeding displays are different, but both birds use high flying and vigorous dives, exposing their pale underparts to produce a signal that is visible over a wide area in the open country they inhabit.

Both species nest on the ground, often within patches of tall heather, although the owl makes do with a simple scrape in which to lay its eggs. A sizeable clutch can be laid, up to as many as nine eggs (even more in the owl), and this allows them to take full advantage of years when vole cycles are at their peak and food is abundant. The young develop very quickly, leaving the nest to hide in vegetation close by before they are able to fly, a shared adaptation that helps minimise losses to ground predators.

One key difference is that, in contrast to the harrier, male and female Short-eared Owls share a similar plumage, though the males tend to be somewhat lighter and less heavily marked. Nevertheless, the owl's overall pattern and colouring, darker above than below, is rather similar to a female Hen Harrier, including the band of more heavily marked feathers towards the upper breast. Finally, and perhaps most striking of all, both birds have the distinctive facial disc of stiff feathers which helps them to pinpoint prey by sound as well as by sight.

Whiling away the time with the Short-eared Owls. Before looking for the owls, we had a little wander around the top of the pass. The air was clear, giving splendid views across the moors and on to distant towns. It was wonderful to watch Curlews picking through the grass tussocks, while Swallows swept by snatching at insects, and several Red Grouse sneaked along the ridge tops, keeping an eye on us. Among the moths was a Latticed Heath, a typical species of the moors. A Mountain Hare grazed in the distance.

In place looking over a wild, beautiful moorland pass, mainly made up of deep grasses, heather, rocky crags and drystone walls. At about 17.40 – first owl, hunting near the top of the pass. Several plunges into the long grass. A second owl, much closer – hunting, plunging and landing on small boulders. Staring around, and then more hunting. Golden Plover calling from the distant crag – then take-off, calling in alarm. The near Short-eared Owl belts across the road and starts to gain height. Two Buzzards are flying along the crag. Dog fight. The Short-eared Owl ascends above them, turns and with powerful, straight wingbeats hammers at one of the Buzzards with great speed and agility. Appearing quite small against the bulk of the Buzzard. The Short-eared Owl repeats this manoeuvre several times, going for whichever Buzzard is in range. Sometimes forcing them to flip over and defend themselves with their talons. A pair of Curlew join in to see them off, not in the dogfight, but with vocal support. Job done, the Short-eared Owl lands on a stone wall, looking a bit puffed, but sporting that fierce stare that only owls can muster. The Golden Plovers are still flying about calling. A third Short-eared Owl appears and, as we left, a fourth flew over our car from a conifer plantation.

20/5/19
DRAMATIC SKY.

CUPOLA

SEPTEMBER

IN THE SAME WAY THAT MARCH marks the transition from winter to spring and the breeding season to come, so September sees the annual cycle swing back in the opposite direction. The young harriers are now fully independent of their parents and the nesting area where they were reared. Breeding is finally over for another year. Some birds might still be seen in the breeding areas, and a few individuals, especially the larger females, remain there for the winter if conditions do not become too hostile and the food supply holds up. But for the majority of birds, the search begins for a more benign place to spend the winter. They seek out lower ground, often near to the coast, and some end up hundreds of kilometres away from their breeding site.

Wanderlust – movements and migration

The pattern of movements of Hen Harriers in Britain defies easy description. Further north and east in northern Europe and Asia, where winters are severe, the picture is more straightforward. Here, the species can be described as a true migrant. The breeding grounds are abandoned and the birds head south to places with a gentler climate and a more reliable food supply where they can see out the winter. They behave in a similar fashion to more familiar migrants such as Honey-buzzards, Ospreys and Hobbies. In British Hen Harriers things are rather more complicated, though we are now starting to understand the patterns of movements more fully. The recovery of birds ringed as nestlings and found dead in winter provides some information. And a long-running RSPB study

Taking a relaxing stroll on a long, sandy beach, away from the moors. A harrier comes in off the sea and flies down the shore. Left: A Common Sandpiper bobbing its tail (as they do). It too will soon be heading south, though it will probably travel further than the harriers that have been breeding on the adjacent moor.

A ringtail harrier had stopped off for a while in the Alver Valley, Gosport. Not wanting to miss out and with a scheduled meeting in the vicinity, we gave ourselves a little time to catch up with it. An hour later we had been entertained by Stonechats and a Dartford Warbler, but no harrier. Time to go. And as if on cue the harrier appeared and started to quarter the scrub, with two attendant Magpies. We tucked ourselves into a hedge and watched this superb bird perform. At times coming within a few yards. Eventually the bird plunged into the rank grass, rose up again carrying what looked like a vole and dropped back into the grass to take its tea. Our cue to get back to work. The client was very understanding about our tardiness – after all, he was the one who told us about the bird.

that fitted wing-tags to over 1,600 nestlings in Scotland and Wales over many years allowed interesting movements to be detected when the tagged birds were re-sighted. In recent years, the use of radio and satellite tags has added texture to our understanding, revealing movements day by day as birds start to move away from their breeding sites.

As we have seen, some birds, especially females, remain on, or close to, the breeding grounds throughout the year. Even in the far north, in Orkney and in the Hebrides, where the warming Gulf Stream helps reduce the severity of winter, many birds remain year-round. For those that do make long-distance movements, tracking has shown that most do not make one direct movement between their nest site and the place they will spend the winter. Instead, they roam widely in the autumn, often covering huge distances and spending a few days or weeks at a number of different sites. Movements can be in any direction including to the north of the breeding site. These birds may simply be searching for a good place to settle for the winter but, more likely, there is a genuine element of exploration. They are learning about different areas, discovering if other harriers are present and how easy it is to find food in different types of landscape. Perhaps they are also checking out the lie of the land with one eye on where they might settle to breed in future. Surviving winter is one thing, but it will all be for nothing if they don't find a good place to nest and rear their own young. They will have a head start if they are already familiar with a number of potential options come the following spring.

A further complication is the exchange of harriers between Britain and the Continent. At least a small number of birds raised in Britain end up travelling overseas to spend the winter, as shown by the pioneering early work based on ringing nestlings in Orkney. Birds from this study were recovered from as far afield as Norway, Denmark, the Netherlands, Germany, eastern England and the south-west of Ireland. Some birds from mainland Scotland move south-west into Ireland or south through the Pennines of northern England, heading down into the south of England or across to Wales. A few, almost all males, then cross the channel en route to wintering areas in southern France or Iberia.

Making the journey into Britain are birds from northern and eastern Europe that head south and west to winter mostly in the southern half of England. Because the Hen Harrier has a light wing-loading and is capable of covering large distances using soaring and flapping flight, it tends not to gather at well-known raptor migration hotspots. Other raptors congregate at sites such as Falsterbo, in southern Sweden, in order to cross the sea at its narrowest point, avoiding long and gruelling flights over open water. Migration for the Hen Harrier is on a broader front and so tends to go largely unnoticed. But ring recoveries and sightings of tagged birds have shown that individuals from nests in Finland, Sweden and the Netherlands can cross the sea to winter in southern and eastern England. If the weather takes a turn for the worse then the generally milder conditions in England become even more attractive and there may be a 'cold-weather influx' of birds from further north and east, seeking to escape the tough conditions.

Even in this age of technology, with satellite tags deployed to track movements, the humble ring still has something to offer. Rings are cheap and so can be fitted to many birds. They last a lifetime (with no batteries to worry about), and reports of ringed birds found dead provide insights into movements, survival rates and causes of mortality.

4/2/09

OCTOBER

OCTOBER SEES A CONTINUED MOVEMENT of Hen Harriers towards their wintering grounds across Britain. By the end of the month the majority of birds will have completed the transition. While they spend their days hunting across the open fields and marshes, every afternoon they come together with others of their own kind to spend the night. The communal roosts, for which the Hen Harrier is so well known, are now reforming and traditional sites that may have been used by generations of wintering birds are occupied once again. Birdwatchers across the country have a chance to renew their acquaintance with an old friend.

The benefits of communal roosting

Many birds, from a variety of different families, roost communally and there is no lack of theories as to why they do so. It may be because good places to spend the night are in short supply and so all the birds in an area home in on the only suitable locations. Huddling together can help to retain warmth and improve the chances of survival on the coldest of nights, as with Wrens cramming themselves into a nest box or tree-hole. Gathering together in numbers can help with the early detection of predators and it offers the opportunity to share information on good places to find food, either directly, or through birds following each other when they fly out of the roost the next morning. There may also be a wider, social element at play, with roosts offering a place where males and females can meet up and assess each other as potential mates for the breeding season to come. The reasons for roosting communally vary from species to species and they are not mutually exclusive. There are likely to be a number of advantages that result from a behaviour that has become such an integral part of the lifestyle of many birds.

So, why does the Hen Harrier roost communally in winter? In truth, this is a difficult question to answer and while it is likely that several different reasons apply, the relative importance of each is still far from clear. The one benefit that

As the harriers ready themselves to roost in the reedbed, a Bittern lands in an open pool and begins to hunt – neck stretched out.

can safely be ruled out is that of helping to maintain warmth. Hen Harriers do not huddle together when they are roosting. Each individual settles on its own small patch of ground, generally a metre or more away from the other birds nearby. The dense vegetation helps provide warmth and shelter but the presence of the other birds does not add to the effect.

Predator detection and defence

A bird as large as the Hen Harrier roosting on the ground offers the chance of a sizeable meal for a patrolling Fox. The fact that roosting takes place at traditional sites must add to the risk, as predators can learn the places that are worth checking regularly. Hen Harriers have excellent eyesight and hearing but a predator could approach from any direction, its shape concealed by the thick vegetation and any sounds masked by a breeze rustling through the dry grasses. In this situation, the more pairs of eyes and ears the better. The Hen Harrier's excellent hearing is usually thought of as an adaptation for hunting but it must also be useful for sensing danger when roosting on the ground. If just one harrier hears (or sees) something untoward then it will fly up, raising the alarm for all the birds present. Roosting together also provides the opportunity for group defence. A number of harriers diving down repeatedly at a Fox, strafing it from all directions, might give it cause to think twice about paying another visit to the same site. Aerial predators too may be mobbed with the same intent, to help encourage them on their way.

While such behaviour helps reduce the risk of predation, it is no guarantee of survival. When fieldworkers visit roosts during the day they occasionally find a sad scatter of harrier feathers on top of a flattened patch of vegetation. They mark the loss of one of the roost members, and, close by, a tell-tale Fox dropping gives away the culprit. It is also worth reinforcing a point made earlier: the communal roosting habit, centred on traditional sites, has increased the vulnerability of the bird to human persecution because of its regular, predictable appearance in the same locations. Evolution works over vast timescales to refine beneficial behaviours in a natural situation. It is far less effective at responding to the rapid changes brought about by humans.

Sharing information about food

Research has shown that this is important in some communally roosting birds of prey, though it is best known in scavengers that feed on animal carcasses. This type of food lasts for several days and individuals can learn about its location by following well-fed birds from the roost site early in the day. Sharing information in this way is less likely to be important for an active hunter like the Hen Harrier. Nevertheless, it's possible that birds that have struggled to hunt successfully

The bulging crop of this recent arrival at the roost shows that it has had success in its hunting efforts during the day.

may try to learn about areas with a richer food supply by following the more successful individuals. This is very hard to prove, but observations reveal that several harriers sometimes head off at the same time and in the same direction when they leave the roost. It seems unlikely that this is mere coincidence. Are they all heading to areas with which they are already familiar from their own direct experience of hunting? Or might a less experienced bird be seeking to learn something new by following one of its fellow roost mates? Wintering birds new to an area may also learn about the location of the roosts themselves by observing and then following more experienced birds in the late afternoon.

Socialising

This is another facet of bird behaviour that is notoriously difficult for humans to understand through observation alone. We marvel at the aerial displays of harriers as they wheel and cavort above the roost, diving at each other in mock fights and sometimes displacing each other as one flies to a spot where another has already settled. But we can only guess at the meaning of these interactions and whether they result in any long-term bonds being formed. Perhaps one bird displacing another is nothing more than an attempt to claim a highly valued roosting spot. But when a female displaces a male this mirrors a behaviour that is sometimes seen at nest sites involving birds that are known to be paired. Observers at roosts may see birds that appear, from their close associations, to have paired up, though it is impossible to say whether such pairs were formed at the roost site or are long established.

HEN, MARSH HARRIER
MERLIN, WHOOPER SWAN
BARN OWL - BUT NO
HORSEY MERE
FEB '88
HARRIERS DRIFTING TOWARDS THE ROOST.

Opposite: Brograve Mill, Horsey Mere, Norfolk. At the end of the day, Marsh and Hen Harriers begin to gather, along with Merlins. Whooper Swans rest on the mere and a Barn Owl starts its evening shift. No Cranes though.

It seems likely that unpaired birds interact with potential future mates at communal roost sites. The spectacular aerial sparring and play-flighting might be part of a process by which males and females are weighing each other up as future partners.

NOVEMBER

WINTER FOOD

A FAR WIDER RANGE OF PREY SPECIES are taken in winter than in the breeding season. This is because, free from the constraints of a fixed nest site, harriers hunt over a variety of different habitats, each supporting a diverse community of birds and mammals. Small and medium-sized animals dominate the diet, and, as in the breeding season, the female is able to subdue larger and heavier prey. Almost any species could be on the menu, provided it is small enough to be tackled and spends time in the habitats hunted over by harriers. Numerous studies of winter diet in Britain have resulted in a list of dozens of different prey species.

The Hen Harrier might be considered as a specialist in its hunting technique of low flights and rapid darts at prey surprised on the ground, or flying up too late to avoid being caught. But it is a generalist in terms of the huge range of animals that it is able to exploit. Even so, within the broad range of prey taken, there are key species in each area that are encountered most frequently and are the easiest to catch. It is these that the harriers rely on most heavily.

It is difficult to work out what Hen Harriers are feeding on by direct observation alone. Kills are rarely witnessed and when they are seen, the action is fast and furious, and it is rarely possible to identify the prey involved. Studies of diet in winter have mostly been based on pellets found at communal roost sites which contain the undigested remains of fur, feathers and even bird beaks. One detailed study involved hundreds of pellets collected from roosts in north Norfolk, the New Forest in Hampshire and Wicken Fen in Cambridgeshire. This found that while a wide variety of different birds were predated, those often encountered out in the open, away from dense cover, predominated. The most common birds were Skylark, Reed Bunting, Yellowhammer, Greenfinch, Linnet and Dunnock. The last species was perhaps the most unexpected, as Dunnocks are often thought of as birds that frequent bushes in woodland or scrub. But Dunnocks will also venture into open fields, where they become vulnerable to harriers. The study went a step further by analysing the weed seeds found in the harrier pellets, seeds that were consumed incidentally when seed-eating birds were predated. In effect, the harriers were co-opted to investigate which weeds were important in the diet of some of our common farmland birds. This novel and innovative approach highlights the importance of a complete food chain for our top predators. If there are fewer agricultural weeds because of intensive farming practices, there will be fewer seed-eating songbirds and, ultimately, fewer wintering harriers.

Another study compared the contents of pellets from Breckland, on the Norfolk/Suffolk boundary, and the New Forest. The Breckland birds hunted mainly over the surrounding arable farmland, and Skylark, Dunnock and Greenfinch were the most important species. In the New Forest, there was a mixed picture, with the diet influenced by the main habitats surrounding the roosts. Birds from roosts close to farmland took mainly Skylarks, with Wrens and Linnets the next most frequent. In the south of the New Forest, where heathland is a more dominant habitat, they fed mostly on Wrens, Meadow Pipits and even Dartford Warblers, birds typically common on heathland. In late winter, lagomorphs (probably mostly Rabbits) became more important in the diet.

Pellets from an upland roost site in Strathclyde, southern Scotland, provided very different results from those in lowland England. Here, mammals were dominant in the diet, made up of Rabbits and hares (whether these were Brown or Mountain Hares could not be distinguished) as well as small mammals such as voles. Birds were also taken regularly, with gamebirds, notably Red Grouse, the most important group, especially for females. Few small songbirds were taken, reflecting their movement away from the uplands to lower ground during the winter. Mammals were also found to be important in Orkney, with voles outnumbering Rabbits. Songbirds were predated regularly in Orkney but, in contrast to southern England, the Starling was the most important species, reflecting its abundance in the pastures over which the harriers hunted.

There is good evidence that male harriers take more small birds than females, as would be expected from their smaller size and greater manoeuvrability. Pellets from roosts in Scotland, for example, tended to contain more songbird remains when they held a higher proportion of male birds. Occasionally, the remains of birds as large as an adult Pheasant or wildfowl of a similar size are found in harrier pellets. It is likely that these have been scavenged rather than killed, something that most birds of prey will resort to if they are desperate for food and live prey has been hard to come by. But it is also possible that the harriers took advantage of birds that were injured or sick and so less well able to defend themselves.

Daily food requirements

As a rough estimate, a medium-sized bird such as the Hen Harrier will need to find prey amounting to between 10 and 15 per cent of its own body weight, on average, every day. The exact amount will vary depending on the amount of activity undertaken and the temperature. More is needed in cold weather when birds must burn additional fat in order to maintain their body temperature. And, if food becomes hard to find, they must spend more time hunting, which also uses up valuable energy. No doubt birds can survive for a few days if hunting is unsuccessful, but they will quickly begin to lose condition.

Using a figure of 12.5 per cent as an estimate, it is interesting to consider what would need to be caught during an average winter's day to keep a male (weighing about 350 g) and a larger female (around 500 g) on an even keel. For a male, at least six Goldcrests would be required, or, alternatively, four Wrens or Blue Tits, three Linnets, two Yellowhammers or House Sparrows, or, if they are lucky enough to catch one, a single Blackbird, Song Thrush or Fieldfare. A female would need to find five Wrens, or one of the larger thrushes to balance the books.

The humble Wren is a tiny bird and yet Hen Harriers consider them worthwhile as prey; their remains are often found in pellets. They are common and, once spotted at close range, perhaps relatively easy to catch given their short wings and low, whirring flight. Dartford Warbler feathers were a surprise finding in pellets from roosts close to heathlands in the New Forest. They too are not the strongest of flyers and so may also be relatively easy to catch compared to other small birds.

THREATS AND SURVIVAL

IT'S A SAD FACT THAT IN BRITAIN TODAY by far the most significant threat to the Hen Harrier's future comes from the activities of humans. An earlier chapter explored the conflict with Red Grouse shooting and the resulting illegal persecution that prevents Hen Harriers from occupying large areas of suitable habitat. Persecution is much less of a problem in lowland wintering areas away from the moors, but it still takes place, and as the Hen Harrier is under such intense pressure in the uplands, the loss of even a small number in winter only adds to its problems. One infamous case showed that almost nowhere is safe. It occurred on the Queen's Royal estate at Sandringham in Norfolk and involved the shooting of two Hen Harriers in November 2007, witnessed by a warden on the neighbouring nature reserve. Prince Harry was in the area at the time and was interviewed by the police, leading to some eye-opening newspaper headlines, but no further action was taken; perhaps his memoir will reveal more about this incident.

Land management

The way humans manage the land has a direct effect on harriers. Initially, human impacts would have been positive, as forests were cleared and more open country created, over which harriers could hunt. But more recently, land management has intensified, resulting in habitats becoming less favourable for many species. Much of the uplands are heavily grazed and patches of vegetation are burnt in order to promote new growth for livestock or grouse. When the vegetation becomes too short and unstructured it provides less cover for the harrier to nest and supports fewer of the voles and songbirds upon which it depends for food. The effects of this were seen clearly in Orkney, where a once-strong harrier population declined due to a reduction in the food supply caused by heavy grazing. Thankfully, numbers have now recovered following reductions in grazing pressure and a resulting increase in the amount of rough grassland. Recently, another potential threat has come to light in Orkney. Stoats, not native to the islands, have been introduced by humans. If they become well established, they could have an impact on vole populations, so reducing food availability, as well as becoming a significant predator (in the absence of Foxes) at Hen Harrier nests. A multi-million-pound eradication programme, using specially trained detection dogs, is now under way to try to eliminate this new threat.

Hen Harrier populations in Northern Ireland, Wales and the Isle of Man have all declined in recent years, with a variety of factors implicated. In Northern Ireland, direct persecution is not thought to be a major problem, so the focus has been on habitat degradation and a resulting decline in food availability. Wildfires and peat extraction using heavy machinery have damaged some areas of moorland, reducing the quality of the habitat and the food supply. And it is likely that forest management work and increased disturbance from recreational use have had adverse effects on birds breeding within plantations. In Wales it has been suggested that recent cold springs may have knocked back the population, and reductions in food availability due to land drainage and overgrazing are also probably contributory factors. On the Isle of Man, where the population has now stabilised, previous declines were difficult to explain. It might be that declines on mainland Britain have reduced the number of immigrants reaching the island, something that could also be a factor affecting harriers in Northern Ireland and Wales.

Away from the breeding areas, lowland farmland is also managed in an increasingly intensive way and this has led to well-publicised declines in small farmland birds. Once common species such as the Tree Sparrow and Corn Bunting have become much harder to find, and even common birds, including Skylarks, Linnets and Yellowhammers, are far less abundant than they once were. Hen Harriers wintering on farmland must surely find it more challenging than they once did to secure enough food to keep them going through the winter.

Poisoning

The use of agricultural pesticides is one aspect of intensification that has impacted songbird populations by reducing their food supply, thus limiting the availability of prey species for harriers. In the past, certain pesticides have also had direct effects on songbirds and their predators. Organochlorine pesticides such as DDT were widely used into the 1960s. They poisoned small birds that fed on the treated cereal seeds and, in turn, the predators feeding on these birds were poisoned. Hen Harriers were among the victims, which is unsurprising given their occurrence in winter on lowland farmland and their propensity to feed on farmland birds. Elsewhere in Europe, pesticides based on mercury have been implicated in Hen Harrier deaths. All these pesticides were banned many years ago, when their impacts on wildlife were finally revealed, but new products are emerging all the time. In recent years, neonicotinoids have been much in the news for their harmful effects on pollinating insects and likely adverse effects on birds. Vigilance is needed to ensure that we do not end up repeating the mistakes of the past. Monitoring the health of top predators, including the Hen Harrier, is an important part of the picture.

The use of lead in ammunition by hunters is a well-documented threat to birds of prey, including the Red Kite and the White-tailed Eagle. Lead fragments are ingested with prey and then dissolve in the acidic conditions in the stomach, releasing the poison with potentially lethal results. This is also likely to cause problems for the Hen Harrier, although this has not been well studied. Harriers will occasionally scavenge at carcasses that may have been shot, and the larger females regularly take gamebirds such as Red Grouse, especially if they have been injured and are easy to catch. This draws them to the very birds that present the greatest risk of lead poisoning.

As long as lead is used in ammunition for shooting gamebirds, in preference to the safer alternatives that are readily available, poisoning will remain a threat to Hen Harriers, and will continue to kill other predators and scavengers.

A male returns to the nest, against the patchwork landscape of a driven grouse moor.

Collisions and disturbance

Hen Harriers are occasionally killed when they collide with artificial structures. As a bird that has evolved in open landscapes, with few natural obstacles, it is not well adapted to cope with the arrival of novel structures in its environment. Overhead wires present a threat as they are difficult for birds to see, especially when they are scanning the ground beneath them for prey, or are in hot pursuit of a fleeing bird. The occasional Hen Harrier is killed by collision with road traffic and there is at least one record of a bird being killed by a train, perhaps taken by surprise when scavenging on the tracks. The increase in wind turbines, especially in the uplands, is an additional threat. Because harriers tend to hunt low to the ground, they will usually remain below the level of the turbine blades. But in the breeding areas they fly much higher during their displays, when they

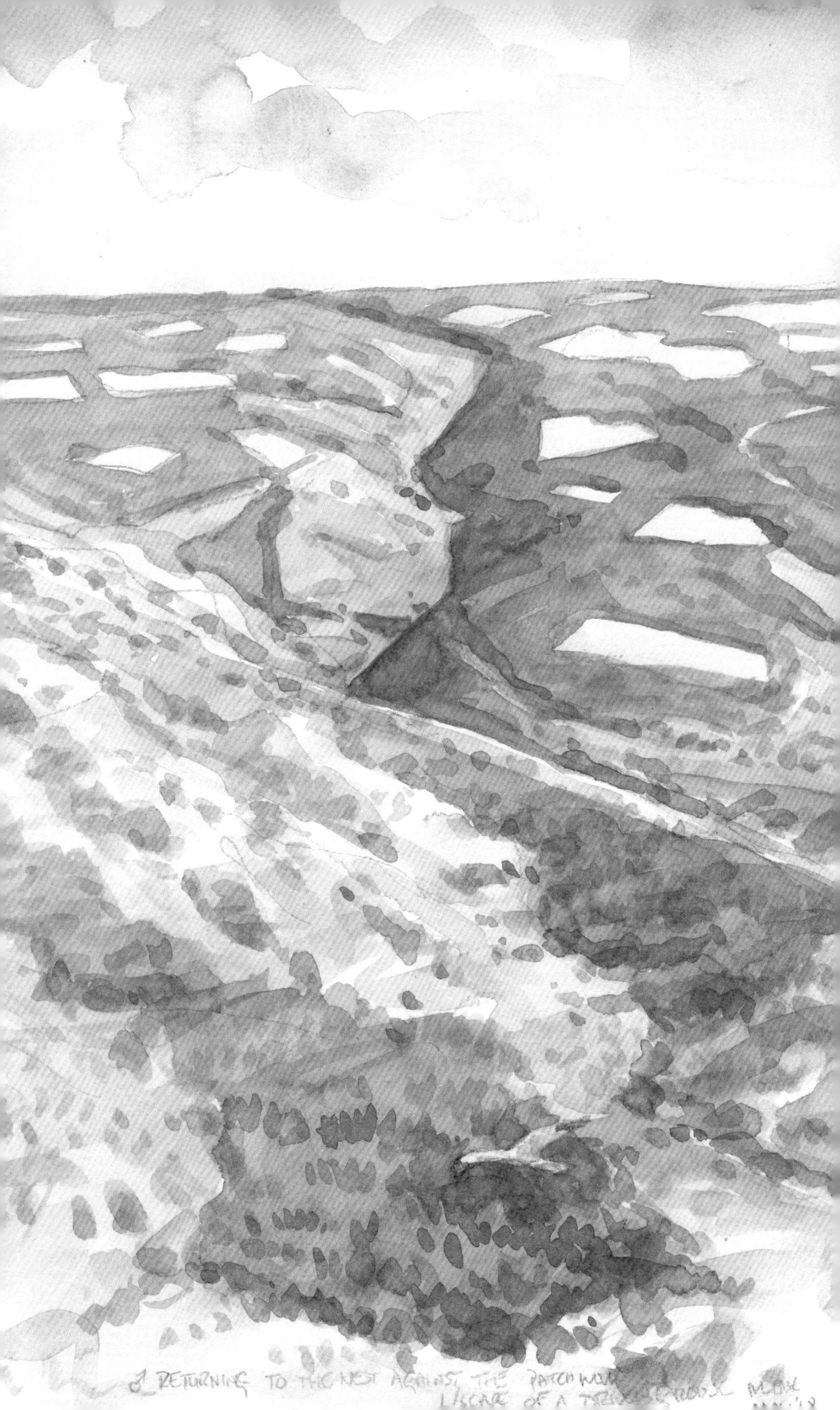
♂ RETURNING TO THE NEST AGAINST THE PATCHWORK

are concentrating on putting on a good show to impress potential partners. A few deaths have been recorded, showing that the risk is real, though it appears from early studies that Hen Harriers are not as vulnerable as birds that spend more of their time flying high above the ground.

Accidental disturbance by humans, both at nest sites and communal winter roosts, is a cause for concern, especially as our remaining wild spaces become smaller and subject to ever greater pressure from visitors. Individual birds vary in their response to disturbance. Many are able to adapt to some extent, especially if the disturbance is predictable, such as along a well-used public footpath. But others are less tolerant, and unpredictable human activity or intrusions into areas that usually see few human visitors can lead to problems. Nests may be left unattended and the young then become vulnerable to predators or chilling in poor weather.

At winter roosts there are no young to worry about, but if birds are disturbed repeatedly they have to expend extra energy in flying to an alternative roosting site and they may be forced to spend the night somewhere less suited to roosting – with a lack of shelter from the elements or a greater threat from ground predators perhaps. Ultimately, this could result in lower survival rates.

Deaths from natural causes

It is often difficult to separate natural causes of death from those associated with humans. In our busy modern landscapes, the two are closely linked. For instance, a bird that is close to starvation is likely to take more risks as it becomes increasingly desperate to find food. It must hunt for longer periods and will be less alert to its surroundings, making it more vulnerable to illegal persecution and to collisions. Even starvation and disease, both 'natural' causes of death, are

January 2020. As a birthday treat we nipped over to Dartmoor to watch roosting harriers somewhere with a different landscape from the New Forest. Well, the weather was certainly interesting – sunshine, to snow, to freezing winds, to gloom, then back to sunshine and wind. We had several views of harriers. The first, two ringtails, which we nearly missed because we were distracted by a herd of Red Deer ambling along the slope of the hillside. Others came and went (the same birds?), but none went into the place we felt looked most promising as a roosting area – a bog with tussocky islands. Heavy rain. The following evening and slightly different circumstances. We expected the strong winds, but not being blocked in by four-by-fours with angry-faced drivers. We had been here before, it could only mean one thing, a foxhunt and we were deemed potential hunt saboteurs. Nice, no human rights infringed here.

With little chance of seeing a harrier, or of leaving for that matter, we sat back with a mug of coffee and watched events unfold. As far as we could make out the dogs were in control by being out of control, racing around, following wherever the scent took them – straight through the potential roost site at one point. Hunt support was in chaos, too – lots of shouting at dogs. People in red jackets on horses always about ten minutes behind the action. Meanwhile a Fox sauntered off into the sunset. Just a thought: isn't foxhunting banned in the UK? Funny old way to celebrate my 60th.

influenced by the way that we manage the habitats upon which harriers depend, and the amount of food that is available to them. A bird in poor condition because its food supply has been limited by intensive farming is more likely to succumb to illness or predation than a healthy, well-fed individual.

A wide range of aerial and ground predators take Hen Harrier nestlings, and the occasional incubating adult when they get the opportunity. Recently fledged young are also vulnerable while they are developing their hunting skills and are occasionally killed by Peregrines, Golden Eagles, Goshawks or even Eagle Owls where this species has become established in the wild. Adults are occasionally killed by large birds of prey, and the Golden Eagle may even have an impact on Hen Harrier breeding densities, by dissuading them from settling in an area. Indeed, it has been suggested that if Golden Eagles were allowed to spread onto grouse moors through a relaxation in illegal persecution, this might help alleviate the conflict between Hen Harriers and Red Grouse. The Peregrine too is large and powerful enough to be viewed as a threat by adult harriers – as evidenced by their reaction when one flies close to their roost site (see *December*). Large ground predators, especially Foxes, pose a threat at communal winter roost sites as discussed earlier. They may also occasionally manage to surprise a bird that is feeding or investigating potential prey on the ground. The risk is greatest in the rank vegetation that harriers favour for hunting, when sightlines are obscured. With potential threats from the air and on the ground, it is as well to be on high alert at all times.

Life expectancy

Despite the complex range of factors with the potential to impact on Hen Harriers in Britain and Ireland, one thing is clear: many birds are killed illegally before their time and the overall population is held well below the level that the environment can support. If this one problem alone could be tackled, the population would recover rapidly and provide recruits to help recolonise the places where harriers are currently scarce or absent. This was demonstrated very effectively during the Langholm Moor conservation project in southern Scotland. Agreement was reached with the landowner that illegal persecution of harriers should stop for the duration of the project. Freed from this constraint, the population increased dramatically from 2 to 20 pairs in just five years.

As with most animals, young and inexperienced Hen Harriers suffer far greater losses than adults. Recently fledged young have perhaps no more than a 20 per cent chance of surviving for long enough to make their first breeding attempt, a figure that is lower than it should be due to illegal persecution. The odds are very much against them, especially up on the grouse moors. For those that make it through their first 12 months, the situation improves. They are now more streetwise. And they have a track record of being able to find enough food and avoid predators, including humans. Threats remain but annual survival rates increase to around 70–80 per cent, again varying between areas and influenced strongly by persecution. An adult bird might expect, on average, to live another four or five years. The lucky ones, dwelling away from persecution hotspots, are capable of living for much longer and the current record, based on the recovery of a ringed bird, is 15 years and 9 months. It would be fascinating to know how many young harriers that particular individual helped to rear during its long life.

DECEMBER

DAY-LENGTH CONTINUES TO SHORTEN as the year draws to a close, reaching its nadir a few days before Christmas. For harriers struggling to find food, the time available for hunting is limited. To survive they must make the most of the short days and spells of settled weather. The low light levels of midwinter can affect people adversely too, though these days it is our sense of wellbeing that is compromised rather than our hunting ability or food supply. If you find that the dull conditions and short days are getting you down, what better way to revive the spirits than an afternoon walk, through crisp winter air, to a Hen Harrier roost. Here, there is the chance to witness behaviour that has been re-enacted over countless generations, at traditional sites scattered all across Britain. These places provide a reminder that there is so much more to this world than the daily trials and tribulations of human society.

Visiting a communal roost

Communal gatherings of birds as they arrive at their roost site provide some of the most impressive wildlife-watching opportunities that Britain has to offer. Starling murmurations provide the most familiar example, and involve the largest numbers, with tens or even hundreds of thousands of birds performing their aerial acrobatics before descending into their roost for the night. Red Kites and Marsh Harriers both roost communally in winter and, if luck is with you, it's possible to see over a hundred of each at favoured sites. In India, harrier roosts may involve thousands of birds, mostly Montagu's and Pallid Harriers but also a few Marsh and Hen Harriers, if you are able to pick them out amid the mêlée of aerial activity.

The average Hen Harrier roost in Britain tends to be rather more understated, with the number of birds usually in single figures. Occasionally, a well-used roost might hold up to 20 or 30 birds, and until relatively recently a famous roost at the Curraghs on the Isle of Man supported up to 160 birds on the best nights, although this site is now little used. Yet even a small harrier roost more than repays the effort of a winter visit. There is something appealing about the ritual of it. It starts with arriving at your chosen site in mid-afternoon, while the

Bulgaria. We tootled down to the Durankulak Lake to watch Red-breasted Geese flight in. As the light dimmed, skeins of geese bundled their way onto the lake. This would provide a fitting conclusion to any day, but this scene was further enriched by an aerial performance of some grace, involving a cast of Hen and Marsh Harriers, Bitterns, Starlings, Pygmy Cormorants and a White-tailed Eagle. Plus a guest star – a Saker Falcon toying with the harriers and then attacking a Magpie that thought it a good idea to taunt the falcon. Such a privilege to witness. Back then to the warmth of the lodge to relive the day with our friends and a local cognac or three.

light is still good and the birds are away on their hunting grounds, nowhere to be seen. There is the building anticipation as the horizon is scanned, and that initial flush of excitement as the first harrier of the evening shows itself above the skyline. And then (hopefully) comes the flurry of activity as more birds appear in the distance and begin to track steadily towards the roost. As with other forms of birdwatching, part of the appeal lies in a sense of detachment and escape from a world dominated by humanity – a glimpse into the barely knowable world of another species. Donald Watson loved visiting his local roosts in Dumfries and Galloway, and he clearly shared this view:

> *For an hour or so towards the end of a short winter's day, a watcher can break all ties with the world of human affairs and enter another ... which is the world of wildlife. I often wondered how far back in the distant past harriers had first discovered and made use of the flow for roosting. There was something about the slow pageant of their gatherings which contributed to a sense of timelessness. Watching the final moments of the harriers' day seemed to underline how little I knew of their total activity.*

As at Starling roosts, some of the most impressive aerial displays are prompted by the appearance of predators. Harriers have been observed rising into the air above their roost, trying to get above an intruding Peregrine so that it cannot dive down onto them. The bolder individuals also look to mob aerial predators to let them know they have been spotted and to attempt to make life uncomfortable for them, hoping this will encourage them to move away. Harriers are normally silent at their roosts but they may use their chittering alarm call when a predator is being mobbed. Ground predators, especially Foxes, are also mobbed, the group acting as one, diving down and alarm calling repeatedly to help persuade the intruder to move on and try its luck elsewhere. Occasionally a human observer is treated in a similar way if watching from a spot too close to where the birds wish to settle: an obvious warning to move away so the harriers don't have to find an alternative site so late in the day.

Influence of weather conditions

There are evenings when each new bird flies directly to the roost and settles down for the night, helping to make counts of attending harriers straightforward. This is more likely when it is raining or when there is little wind, although numbers will probably be lower in such conditions. In some areas, the coldest nights also tend to support fewer birds. It has been suggested that on cold, still afternoons the birds are unable to exploit the wind and so flying is more energetically expensive. They have to weigh up the potential benefits of flying to the roost, set against the effort that would be required to get there. In unfavourable conditions they may settle for a smaller roost closer to their hunting grounds or even a night spent roosting on their own. An alternative suggestion is that the shelter provided by a high-quality roost site is more beneficial in strong winds and so more birds tend to arrive in such conditions. Donald Watson suggested that some of the waterlogged sites used by harriers close to his home in Dumfries and Galloway may be used less in frosty conditions because they become less secure. Ground predators could, he thought, use the ice to reach previously inaccessible patches of ground. There is also a risk that the feet of the roosting birds could become trapped in a layer of ice if wet ground freezes hard overnight.

When there is more of a breeze, flight is easier and the birds tend to spend more of their time in the air, wheeling around over the roost area, cutting across the flightpath of other birds and playfully diving at each other. Just before the birds finally settle for the night there may be a short period when they all seem to be in the air together, though it is impossible to be sure that some do not remain concealed on the ground. Once they have settled, a sudden unexpected disturbance sometimes causes them all to rise up into the air and that, too, provides an opportunity to make a count. Part of the ritual of roost watching is in trying to log the numbers of individuals present. This is helped by the fact that adult males are easy to separate from the juveniles and females. But it can still be challenging. It is especially difficult to distinguish between new arrivals and birds that are already present but fly up after settling in one place only for a few minutes of rest.

Another option for the harrier enthusiast is to visit a roost site in the morning to watch the birds leave. This requires an early alarm call because birds tend to leave very early in the day, often before sunrise, keen to use all the available daylight in their quest for food. They usually exit the roost area quickly, flying low to the ground but sometimes circle up to gain height before heading away in a straight line to a distant hunting area.

Not all roost watches are successful. Some only result in brief, distant records, others a spectacular sunset. But it's always worth that one last look, even long after the sun has disappeared. A glimpse of a male in near dark conditions, making a final flight over the heath as he searches for supper, makes that frothing flagon of ale waiting for you at the local pub taste that little bit better.

The ritual of watching a Hen Harrier roost has been enjoyed by thousands of birdwatchers down the years and contributed much to our understanding of the bird's distribution in winter, and the health of the population. Warm clothing and an eye for detail are essential if you intend to count how many adult males and ringtails are present by the time darkness finally closes over the landscape (see Sources of further information *for details of where to send records).*

Early morning. A male heads purposely away from his roost, low over a frosty plain.

THE WIDER PICTURE

World status

THE HEN HARRIER IS WIDESPREAD IN EUROPE and its breeding range extends to the east in a broad band across Asia all the way to the Pacific coast. Within Europe, it breeds from northern Scandinavia, where it is a summer visitor, south through Central and Eastern Europe, extending as far south and west as northern Spain. In parts of Central and Western Europe it is a resident, as it is in Britain, with at least some of the breeding birds remaining for the winter. In these areas it becomes more widespread in winter as visitors arrive from the north. And across much of Iberia, south-east Europe into Turkey, and in parts of North Africa and the Middle East it is present only in winter.

Although the overall range is very large, the breeding population is declining in most areas for which we have good information. The causes are many and varied

While watching Chamoix in the high-tops of the Central Balkan National Park, Bulgaria, a ringtail Hen Harrier heads northwards against a striking spring mountain scene.

but mostly fall into two broad (and sadly all too familiar) categories: continued persecution by humans and a reduction in food supply and nesting sites due to losses or degradation of habitat. BirdLife International recently estimated the European breeding population at 30,000–54,400 breeding females (of which an estimated 70 per cent were found in Russia) with a tentative estimate for the world population of 176,000–321,000 individuals. That might sound like a healthy population but, to put the numbers in perspective, Britain alone supports 61,500–85,000 pairs of Buzzards, out of a world population estimated at between 2 million and 4 million adult birds. The Hen Harrier has a larger world range than the Buzzard but is spread much more thinly. Overall, the number of Hen Harriers in Europe is well below the level that would be expected from the amount of habitat available, and far short of the population that would have existed before human persecution started to take its relentless toll.

Prospects for the future

The fact that the Hen Harrier breeds across such a vast range offers hope for its long-term future. It is clearly able to adapt to a wide range of climatic conditions and landscapes, making use of many different habitats and taking whatever prey species are abundant locally. While it is a specialist hunter of small mammals and birds, it is also highly flexible in the habitats it exploits and the species that it relies on for food.

An estimated 66 per cent of the world population occurs to the east of Europe. In Russia, in the breeding season, the Hen Harrier can exploit tundra habitats inside the Arctic Circle in the far north, as well as the parched steppe grasslands on the edge of Mongolia, thousands of miles to the south. Wherever it can find open, prey-rich landscapes where it can hunt for food, sufficient vegetation to conceal its nest, and low levels of human persecution, it is likely to thrive. Overall, BirdLife International considers that it is not globally threatened, though closer to home it lists the Hen Harrier as 'near-threatened' on its European red list of birds, due to recent declines. Closer still, here in Britain, its future is far from secure, despite the large areas of available habitat, for reasons that have been explored in this book. It is one of only three birds of prey on the UK's red list of Birds of Conservation Concern. Most raptors have recovered over recent decades thanks to conservation efforts and a reduction in illegal persecution. The Hen Harrier is heading in the opposite direction and in England, despite signs of hope in the last few years, its very existence as a breeding bird hangs precariously in the balance.

Meet Rosie. Back at home in early July it was one of those flat evenings. Rosie's phone 'pings' (not a Covid-19 notification). 'It's from Pat.' 'Rosie, meet Rosie.' And there gazing feistily out from the screen was the little face of a harrier – Rosie. Wiping a couple of tears away, or was it dust? Our minds went back to a chat with Pat in June about what system was used to name the tagged chicks – none. Having decided that 'Dan' didn't really cut the mustard as a good name, we thought no more about it. Of course being a 'Rosie' meant trouble, her satellite tag only transmitted intermittently, resulting in her being declared missing in October 2019.

Ping! Pat again. 'Looks like Rosie's got a fella.' Not missing then. There on the screen, a short film of Rosie dancing with some chap, before going to roost. We approved of him. More dust in the eyes. Not long after, she went off air again: 'Missing, fate unknown at roost.' We would like to think she's just not transmitting again or something natural has happened, perhaps taken by a Goshawk at the roost, rather than the fate that we know, deep down, is most likely. During this work we have asked and been asked what is the solution to the persecution of these birds ... the easy answer is uncomplicated, stop killing them! I wanted to finish off with a well-constructed, cleverly argued and damning piece about: joined-up policies, human greed, arrogance, woeful government complacency, but that would be ending on a rant. Put simply, they deserve better, we owe them better.

The Hen Harrier's fate in the coming years will tell us much about our attitudes towards birds of prey and whether they have changed sufficiently to allow this particular species a continued place in our countryside. Anyone who has delighted in its spectacular skydancing display flights over the moors, watched birds head home towards their communal roost in the depths of winter, or even chanced upon a lone hunting bird, sweeping back and forth across the fields 'as if it were searching for a lost object', will know just how much is at stake.

FURTHER READING

The following list is by no means a comprehensive review of the extensive literature on Hen Harriers but it includes the most important sources of information consulted for this book. Short notes have been added, where relevant, to summarise the type of information provided by each reference. The list includes a mixture of books, reports and scientific papers, all of which should be readily available through booksellers or academic libraries. The summaries of many of the papers listed are available for free online through search engines such as Google Scholar.

Amar, A., Redpath, S. & Thirgood, S. (2003) Evidence for food limitation in the declining Hen Harrier population on the Orkney Islands, Scotland. *Biological Conservation* 111: 377–384.

Arroyo, B., Amar, A., Leckie, F., Buchanan, M., Wilson, J.D. & Redpath, S. (2009) Hunting habitat selection by Hen Harriers on moorland: implications for conservation management. *Biological Conservation* 142: 586–596. https://doi.org/10.1016/j.biocon.2008.11.013

Avery, M. (2015) *Inglorious: Conflict in the Uplands.* Bloomsbury, London. (A compelling account of the impacts of grouse moor management on wildlife, a controversy that sees the Hen Harrier take centre stage.)

Balfour, E. & Cadbury, C.J. (1979) Polygyny, spacing and sex ratio among Hen Harriers *Circus cyaneus* in Orkney, Scotland. *Ornis Scandinavica* 10: 133–141. (Eddie Balfour worked for the RSPB on Orkney and undertook extensive studies of the Hen Harrier in its northern outpost, spanning several decades, until his death in 1974. He contributed to a number of important papers about the Orkney population, including this one, published several years after his death.)

Balfour, E. & Macdonald, M.A. (1970) Food and feeding behaviour of the Hen Harrier in Orkney. *Scottish Birds* 6: 157–166.

Balmer, D.E., Gillings, S., Caffrey, B.J., Swann, R.L., Downie, I.S. & Fuller, R.J. (2013) *Bird Atlas 2007–11: The Breeding and Wintering Birds of Britain and Ireland.* BTO Books, Thetford.

Bijleveld, M. (1974) *Birds of Prey in Europe.* The Macmillan Press, London.

Brown, A. & Grice, P. (2005) *Birds in England.* Poyser, London.

Brown, L. (1976) *Birds of Prey.* Collins (New Naturalist), London.

Carter, I. & Powell, D. (2019) *The Red Kite's Year.* Pelagic Publishing, Exeter. (The first volume in this series, of which this book on the Hen Harrier is the second.)

Clarke, R. (1990) *Harriers of the British Isles.* Shire Natural History, Haverfordwest.

Clarke, R. (1996) *Montagu's Harrier.* Arlequin Press, Chelmsford. (Includes useful information comparing the ecology and behaviour of the Montagu's Harrier with the Hen Harrier and other close relatives.)

Clarke, R., Combridge, M. & Combridge, P. (1997) A comparison of the feeding ecology of wintering Hen Harriers *Circus cyaneus* centred on two heathland areas in England. *Ibis* 139: 4–18.

Clarke, R., Combridge, P. & Middleton, N. (2003) Monitoring the diets of farmland winter seed-eaters through raptor pellet analysis. *British Birds* 96: 360–375.

Cobham, D. (2017) *Bowland Beth: The life of an English Hen Harrier*. William Collins, London. (The late David Cobham wrote this inspiring book shortly before his death. It tells the story of this bird's truncated life, partly imagined and partly based on information obtained from satellite tracking.)

Cocker, M. & Mabey, R. (2005) *Birds Britannica*. Chatto & Windus, London.

Cramp, S. & Simmons, K.E.L. (eds.) (1980) *The Birds of the Western Palearctic, Volume 2*. Oxford University Press, Oxford.

Dobler, G. (2021) Territorial behaviour of the Hen Harrier in winter. *British Birds* 114: 133–147.

Dobson, A.D.M., Clarke, M.L., Kjellen, N. & Clarke, R. (2012) The size and migratory origins of the population of Hen Harriers *Circus cyaneus* wintering in England. *Bird Study* 59: 218–227.

Eaton, M., Aebischer, N., Brown, A., Hearn, R., Lock, L., Musgrove, A., Noble, D., Stroud, D. & Gregory, R. (2015) Birds of Conservation Concern 4: the population status of birds in the UK, Channel Islands and Isle of Man. *British Birds* 108: 708–746.

Elston, D.A., Spezia, L., Baines, D., Redpath, S.M. & Elphick, C. (2014) Working with stakeholders to reduce conflict – modelling the impact of varying Hen Harrier *Circus cyaneus* densities on Red Grouse *Lagopus lagopus*. *Journal of Applied Ecology* 51: 1236–1245.

Etheridge, B., Summers, R.W. & Green, R.E. (1997) The effects of illegal killing and destruction of nests by humans on the population dynamics of the Hen Harrier *Circus cyaneus* in Scotland. *Journal of Applied Ecology* 34: 1081–1105.

Ferguson-Lees, J. & Christie, D.A. (2001) *Raptors of the World*. Christopher Helm, London.

Fielding, A., Haworth, P., Whitfield, P., McLeod, D. & Riley, H. (2011) *A Conservation Framework for Hen Harriers in the United Kingdom*. JNCC Report 441. JNCC, Peterborough.

Forsman, D. (1999) *The Raptors of Europe and the Middle East: A Handbook of Field Identification*. Poyser, London.

Hardy, J., Crick, H.Q.P., Wernham, C.V., Riley, H.T., Etheridge, B. & Thompson, D.B.A. (2009) *Raptors: A Field Guide for Surveys and Monitoring (2nd edn)*. The Stationery Office, Edinburgh. (The definitive handbook for those interested in monitoring our birds of prey.)

Hayhow, D.B., Eaton, M.A., Bladwell, S., Etheridge, B., Ewing, S.R., Ruddock, M., Saunders, R., Sharpe, C., Sim, I.M.W. & Stevenson, A. (2013) The status of the Hen Harrier, *Circus cyaneus*, in the UK and Isle of Man in 2010. *Bird Study* 60: 446–458.

Holloway, S. (1996) *The Historical Atlas of Breeding Birds in Britain and Ireland: 1875–1900*. Poyser, London.

Joint Nature Conservation Committee (JNCC) (2000) *Report of the UK Raptor Working Group*. Bristol: Department for the Environment, Transport and the Regions/Peterborough: JNCC.

Lewis, G. (2017) *Sky Dancer*. Oxford University Press, Oxford. (An engaging and informative novel dealing with the emotive issue of persecution and conflict in the uplands. Aimed at older children but will appeal to all ages interested in the subject.)

Ludwig, S.C., Roos, S., Bubb, D. & Baines, D. (2017) Long-term trends in abundance and breeding success of Red Grouse and Hen Harriers in relation to changing management of a Scottish grouse moor. *Wildlife Biology* 2017: wlb.00246.

Madders, M. (2003) Hen Harrier *Circus cyaneus* foraging activity in relation to habitat and prey. *Bird Study* 50: 55–60.

Marquiss, M. (1980) Habitat and diet of male and female Hen Harriers in Scotland in winter. *British Birds* 73: 555–560.

McMillan, R.L. (2014) Hen Harriers on Skye, 2000–2012: nest failures and predation. *Scottish Birds* 34: 30–39.

Morris, N.G. & Sharpe, C.M. (2021) Birds of Conservation Concern in the Isle of Man 2021. *British Birds* 114: 526–540.

Murgatroyd, M., Redpath, S.M., Murphy, S.G., Douglas, D.J.T., Saunders, R. & Amar, A. (2019) Patterns of satellite tagged Hen Harrier disappearances suggest widespread illegal killing on British grouse moors. *Nature Communications* 10: article number 1094.

Nota, K., Downing, S. & Iyengar, A. (2019) Metabarcoding-based dietary analysis of Hen Harrier (*Circus cyaneus*) in Great Britain using buccal swabs from chicks. *Conservation Genetics* 20: 1389–1404.

O'Donoghue, B.G. (2020) Hen Harrier *Circus cyaneus* ecology and conservation during the non-breeding season in Ireland. *Bird Study* 67: 344–359.

Orchel, J. (1992) *Forest Merlins in Scotland: Their Requirements and Management.* The Hawk and Owl Trust, London.

Picozzi, N. (1978) Dispersion, breeding and prey of the Hen Harrier *Circus cyaneus* in Glen Dye, Kincardineshire. *Ibis* 120: 498–509.

Picozzi, N. (1980) Food, growth, survival and sex ratio of nestling Hen Harriers *Circus c. cyaneus* in Orkney. *Ornis Scandinavica* 11: 1–11.

Picozzi, N. (1984) Breeding biology of polygynous Hen Harriers *Circus c. cyaneus* in Orkney. *Ornis Scandinavica* 15: 1–10.

Potts, G.R. (1998) Global dispersion of nesting Hen Harriers *Circus cyaneus*: implications for grouse moors in the UK. *Ibis* 140: 76–88.

Redpath, S.M., Leckie, F.M., Arroyo, B., Amar, A. & Thirgood, S.J. (2006) Compensating for the costs of polygyny in Hen Harriers *Circus cyaneus*. *Behavioural Ecology and Sociobiology* 60: 386–391.

Redpath, S.M. & Thirgood, S.J. (1997) *Birds of Prey and Red Grouse.* The Stationary Office, London.

Redpath, S.M., Thirgood, S.J. & Leckie, F.M. (2001) Does supplementary feeding reduce predation of Red Grouse by Hen Harrier? *Journal of Applied Ecology* 38: 1157–1168.

Rice, W.R. (1982) Acoustical location of prey by the Marsh Hawk: adaptation to concealed prey. *Auk* 99: 403–413.

RSPB & NCC (1991) *Death by Design: The persecution of birds of prey and owls in the UK 1979–1989.* Royal Society for the Protection of Birds/Nature Conservancy Council, Sandy.

Schipper, W.J.A. (1978) A comparison of breeding ecology in three European harriers (*Circus*). *Ardea* 66: 77–102.

Schipper, W.J.A., Buurma, L.S. & Bossenbroek, P.H. (1975) Comparative study of hunting behaviour of wintering Hen Harriers *Circus cyaneus* and Marsh Harriers *Circus aeruginosus*. *Ardea* 63: 1–29.

Scott, D. (2008) *Harriers: Journeys Around the World.* Tiercel Publishing, Wheathampstead.

Scott, D. (2010) *The Hen Harrier: In the Shadow of Slemish.* Whittles Publishing, Dunbeath. (An accessible account of the Hen Harrier in Northern Ireland, including the discovery of tree-nesting harriers.)

Scottish Natural Heritage (2010) *Diversionary Feeding of Hen Harriers on Grouse Moors – A Practical Guide*. Scottish Natural Heritage, Perth.

Sim, I.M.W., Dillon, I.A., Eaton, M.A., Etheridge, B., Lindley, P., Riley, H., Saunders, R., Sharpe, C. & Tickner, M. (2007) Status of the Hen Harrier *Circus cyaneus* in the UK and Isle of Man in 2004, and a comparison with the 1988/89 and 1998 surveys. *Bird Study* 54: 256–267.

Sim, I.M.W., Gibbons, D.W., Bainbridge, I.P. & Mattingley, W.A. (2001) Status of the Hen Harrier *Circus cyaneus* in the UK and the Isle of Man in 1998. *Bird Study* 48: 341–353.

Simmons, R.E. (2000) *Harriers of the World: Their Behaviour and Ecology*. Oxford University Press, Oxford. (This is an academic book but it is accessible and is full of fascinating information about the ecology of the Hen Harrier.)

Smith, A., Redpath, S. & Campbell, S. (2000) *The Influence of Moorland Management on Grouse and their Predators*. DETR, London.

Thirgood, S.J., Redpath, S.M., Rothery, P. & Aebischer, N.J. (2000) Raptor predation and population limitation in Red Grouse. *Journal of Animal Ecology* 69: 504–516.

Thompson, P.S., Amar, A., Hoccom, D.G., Knott, J. & Wilson, J.D. (2009) Resolving the conflict between driven-grouse shooting and conservation of Hen Harriers. *Journal of Animal Ecology* 46: 950–954.

Watson, D. (1977) *The Hen Harrier*. Poyser, Berkhamstead. (A classic and inspiring monograph on the Hen Harrier by an artist and authority on the species, focusing especially on studies in his home county of Dumfries & Galloway. It has recently been reprinted and so is now, once again, easily available at a reasonable price.)

Watson, D. (2010) *In Search of Harriers: Over the Hills and Far Away*. Langford Press, Peterborough. (A large format book showcasing Donald Watson's artwork, with a short accompanying text, published a few years after the author's death.)

Wernham, C.V., Toms, M.P., Marchant, J.H., Clark, J.A., Siriwardena, G.M. & Baillie, S.R. (eds) (2002) *The Migration Atlas: Movements of the Birds of Britain and Ireland*. Poyser, London.

Wotton, S.R., Bladwell, S., Mattingley, W., Morris, N.G., Raw, D., Ruddock, M., Stevenson, A. & Eaton, M.A. (2018) Status of the Hen Harrier *Circus cyaneus* in the UK and Isle of Man in 2016. *Bird Study* 65: 145–160.

New Forest. Great Grey Shrike. A warming sight on a cold evening while waiting for the harriers to come in.

SOURCES OF FURTHER INFORMATION

Natural England's Hen Harrier Recovery Project: This project has been running since 2002, helping to coordinate monitoring of the small breeding population that survives in northern England. Field staff and volunteers monitor nests and winter roosts, and the project pioneered the use of high-tech satellite tags on Hen Harriers by fitting them to well-grown nestlings. Interim findings were published in a short, accessible glossy report: 'The hen harrier in England' (http://publications.naturalengland.org.uk/publication/83007). More recently, the full results of satellite tracking work have been published in a scientific journal (see Murgatroyd and others in *Further reading*).

Langholm Moor Demonstration Project: A partnership project carried out between 2008 and 2018 to try to help resolve the long-running conflict between raptors and grouse moor management. The project trialled techniques to reduce the impacts on grouse by predation, including diversionary feeding at Hen Harrier nests. Detailed monitoring included the use of nest cameras to help assess breeding productivity and provide information about the food brought back to harrier nests by the adults. For further details and project reports: http://www.langholmproject.com/project.html

The Defra Hen Harrier Action Plan: Sets out the government's plans for restoring Hen Harriers in England including several of the contentious proposals discussed in this book: 'Increasing hen harrier populations in England: action plan' (https://www.gov.uk/government/publications/increasing-hen-harrier-populations-in-england-action-plan).

RSPB's Hen Harrier LIFE + project, 2014–2019: A five-year conservation programme with EU funding combining nest monitoring and protection, satellite tagging, roost counts, investigations work and general awareness raising, focused on northern England, and southern and eastern Scotland: Hen Harrier LIFE project, RSPB (https://www.rspb.org.uk/our-work/conservation/projects/hen-harrier-life/).

'Heads Up for Harriers' Project: A partnership led by Scottish Natural Heritage (now NatureScot), working with landowners, RSPB and the National Wildlife Crime Unit to try to better understand the threats faced by harriers in Scotland and to promote recovery. Includes the use of remote cameras at nest sites and a sightings hotline: 'Heads Up for Harriers', National Biodiversity Network (https://nbn.org.uk/biological-recording-scheme/heads-up-for-harriers/).

Standing up for Nature, written by Mark Avery, and **Raptor Persecution UK** are two essential blogs for following the debate about illegal persecution and the changes needed in order to reduce its impacts in future.

Monitoring Hen Harriers: The Hen Harrier Winter Roost Survey has been running for decades, coordinated by the British Trust for Ornithology (BTO) and the Hawk and Owl Trust. If you are able to undertake regular counts at a roost site then it's worth getting involved in order to ensure that your records contribute to the national picture. See: 'Hen Harrier Winter Roost Survey' (https://hawkandowltrust.org/conservation/projects/hen-harrier-winter-roost-survey). Single roost counts and records of harriers away from roosts can also be reported online through the BTO's BirdTrack website or via the local county bird recorder.

A licence is required to monitor nest sites. Anyone interested in contributing to this work should contact their local raptor group for more information.

Finally, online video-sharing sites such as YouTube are not to be underestimated as a source of entertainment and enlightenment in relation to Hen Harrier behaviour. There are hundreds of videos involving Hen Harriers, including footage from nest cameras, of hunting birds and situations when the tables are turned and other species pose a threat to the Hen Harrier. The closely related Northern Harrier (or Marsh Hawk), considered by some to be the same species as the Hen Harrier, features in many videos from North America, so it is worthwhile searching under these names, too.

SPECIES MENTIONED IN THE TEXT

Birds

Barn Owl *Tyto alba*
Bittern *Botaurus stellaris*
Blackbird *Turdus merula*
Blue Tit *Cyanistes caeruleus*
(Common) Buzzard *Buteo buteo*
Carrion Crow *Corvus corone*
Common Sandpiper *Actitis hypoleucos*
Corn Bunting *Emberiza calandra*
Corncrake *Crex crex*
(Common) Crossbill *Loxia curvirostra*
Cuckoo *Cuculus canorus*
Curlew *Numenius arquata*
Dartford Warbler *Sylvia undata*
Dipper *Cinclus cinclus*
Dunnock *Prunella modularis*
Eagle Owl *Bubo bubo*
Fieldfare *Turdus pilaris*
Goldcrest *Regulus regulus*
Golden Eagle *Aquila chrysaetos*
Golden Plover *Pluvialis apricaria*
Goshawk *Accipiter gentilis*
Great Grey Shrike *Lanius excubitor*
Greenfinch *Carduelis chloris*
Hen Harrier *Circus cyaneus*
Hobby *Falco subbuteo*
Honey-buzzard *Pernis apivorus*
House Sparrow *Passer domesticus*
(Common) Kestrel *Falco tinnunculus*
Lapwing *Vanellus vanellus*
Linnet *Carduelis cannabina*
Long-eared Owl *Asio otus*
Magpie *Pica pica*
Marsh Harrier *Circus aeruginosus*
Meadow Pipit *Anthus pratensis*
Merlin *Falco columbarius*
Montagu's Harrier *Circus pygargus*
Northern Harrier *Circus hudsonius*

Osprey *Pandion haliaetus*
Pallid Harrier *Circus macrourus*
Peregrine Falcon *Falco peregrinus*
(Common) Pheasant *Phasianus colchicus*
Pied Wagtail *Motacilla alba*
Pygmy Cormorant *Phalacrocorax pygmeus*
Raven *Corvus corax*
Red-breasted Goose *Branta ruficollis*
Red Grouse *Lagopus lagopus*
Red Kite *Milvus milvus*
Red-legged Partridge *Alectoris rufa*
(Lesser) Redpoll *Acanthis cabaret*
Reed Bunting *Emberiza schoeniclus*
Ring Ouzel *Turdus torquatus*
Saker Falcon *Falco cherrug*
Short-eared Owl *Asio flammeus*
Siskin *Carduelis spinus*
Skylark *Alauda arvensis*
(Common) Snipe *Gallinago gallinago*
Song Thrush *Turdus philomelos*
Sparrowhawk *Accipiter nisus*
Spotted Harrier *Circus assimilis*
Starling *Sturnus vulgaris*
Stonechat *Saxicola rubicola*
(Barn) Swallow *Hirundo rustica*
Tree Sparrow *Passer montanus*
Water Rail *Rallus aquaticus*
Wheatear *Oenanthe oenanthe*
Whinchat *Saxicola rubetra*
White-tailed Eagle *Haliaeetus albicilla*
Whooper Swan *Cygnus cygnus*
Wigeon *Mareca penelope*
Willow Warbler *Phylloscopus trochilus*
Woodpigeon *Columba palumbus*
Wren *Troglodytes troglodytes*
Yellowhammer *Emberiza citrinella*

Other vertebrates

Adder *Vipera berus*
Badger *Meles meles*
Brown Hare *Lepus europaeus*
Brown Rat *Rattus norvegicus*
Chamoix *Rupicapra rupicapra*
Common Lizard *Zootoca vivipara*
Common Vole *Microtus arvalis*
Fallow Deer *Dama dama*

Field Vole *Microtus agrestis*
(Red) Fox *Vulpes vulpes*
(Common) Frog *Rana temporaria*
Hedgehog *Erinaceus europaeus*
(American) Mink *Neovison vison*
Mountain Hare *Lepus timidus*
Orkney Vole *Microtus arvalis orcadensis*
Pygmy Shrew *Sorex minutus*
Rabbit *Oryctolagus cuniculus*
Red Deer *Cervus elaphus*
Red-necked Wallaby *Notamacropus rufogriseus*
Roe Deer *Capreolus capreolus*
Sheep *Ovis aries*
Slow Worm *Anguis fragilis*
Stoat *Mustela erminea*
Water Vole *Arvicola amphibius*
Wood Mouse *Apodemus sylvaticus*

Invertebrates

Dor Beetle *Geotrupes stercorarius*
Emperor Moth *Saturnia pavonia*
Garden Tiger *Arctia caja*
Green Hairstreak *Callophrys rubi*
Green Tiger Beetle *Cicindela campestris*
Latticed Heath *Chiasmia clathrata*
Northern Eggar *Lasiocampa quercus callunae*
Violet Oil Beetle *Meloe violaceus*

Plants

Bell Heather *Erica cinerea*
Bilberry *Vaccinium myrtillus*
Bog Pimpernel *Anagallis tenella*
Bracken *Pteridium aquilinum*
Climbing Corydalis *Ceratocapnos claviculata*
Cranberry *Vaccinium oxycoccos*
Cross-leaved Heath *Erica tetralix*
Globeflower *Trollius europaeus*
Heath Bedstraw *Galium saxatile*
Heath Spotted-orchid *Dactylorhiza maculata*
(Common) Heather *Calluna vulgaris*
Lesser Spearwort *Ranunculus flammula*
Mountain Pansy *Viola lutea*
Northern Marsh-orchid *Dactylorhiza purpurella*
Thyme-leaved Speedwell *Veronica serpyllifolia*

♀ HEN HARRIER
HUNTING
LATER
WITH PREY

INDEX

Also available from Pelagic

The Red Kite's Year, by Ian Carter and Dan Powell

Rhythms of Nature: Wildlife and Wild Places Between the Moors, by Ian Carter

Human, Nature: A Naturalist's Thoughts on Wildlife and Wild Places, by Ian Carter

Wild Mull: A Natural History of the Island and its People, by Stephen Littlewood and Martin Jones

101 Curious Tales of East African Birds, by Colin Beale

Urban Peregrines, by Ed Drewitt

The Wryneck, by Gerard Gorman

Rebirding: Restoring Britain's Wildlife, by Benedict MacDonald

A Natural History of Insects in 100 Limericks, by Richard A. Jones and Calvin Ure-Jones

Essex Rock: Geology Beneath the Landscape, by Ian Mercer and Ros Mercer

Pollinators and Pollination, by Jeff Ollerton

A Miscellany of Bats, by M. Brock Fenton and Jens Rydell

Ancient Woods, Trees and Forests: Ecology, History and Management, edited by Alper H. Colak, Simay Kırca and Ian D. Rotherham

www.pelagicpublishing.com

♂ HEN HARRIER
LOAFING BIRD